MAI
NORTH AFRICA
and the
MIDDLE EAST

Chris and Tilde Stuart

B L O O M S B U R Y
LONDON • NEW DELHI • NEW YORK • SYDNEY

Bloomsbury Natural History
An imprint of Bloomsbury Publishing Plc

50 Bedford Square   1385 Broadway
London   New York
WC1B 3DP   NY 10018
UK   USA

www.bloomsbury.com

BLOOMSBURY and the Diana logo are trademarks of
Bloomsbury Publishing Plc

First published by New Holland UK Ltd, 2008 as
A Photographic Guide to Mammals of North Africa
This edition first published by Bloomsbury, 2016

British Library Cataloguing-in-Publication Data
A catalogue record for this book is available from the British Library.

Library of Congress Cataloguing-in-Publication data has been applied for.

ISBN: PB: 978-1-4729-3239-6
ePDF: 978-1-4729-3241-9
ePub: 978-1-4729-3240-2

2 4 6 8 10 9 7 5 3

Designed and typeset in UK by Susan McIntyre
Printed in China

To find out more about our authors and books visit www.bloomsbury.com.
Here you will find extracts, author interviews, details of forthcoming events
and the option to sign up for our newsletters.

CONTENTS

SPECIES DESCRIPTIONS

INTRODUCTION

We have long had a fascination with deserts and other dry landscapes, and the many plants and animals that not only survive but also thrive in these harsh environments. The region we cover in this photographic guide contains Earth's largest desert, the Sahara; one of its least explored deserts, the Rub al'Khali; vast areas of arid land in the Arabian Peninsula and extensive tracts of Iran. Here also are high mountain ranges: the Atlas, Zagros, Alborz, Taurus, Pontic and Anatolian. The last contain one of the world's great peaks, Agri Dagi, better known to many as Mt Ararat. The impressive Mt Damavand rises to 5,678 m (18,629 ft) in the Alborz range. Within the Sahara are enormous mountain massifs lying like islands in a great ocean: Ahaggar, Tassili, Tibesti and Aïr. Along the northern fringes of the region sit two of the world's greatest inland seas, the Black and Caspian Seas. The waters of the Atlantic and Indian Oceans, and the Mediterranean and Red Seas, lap the region's shores. The Nile River, Earth's longest river (although some now claim the Amazon for this title), traverses the full length of Egypt and beyond the region to the highlands of Ethiopia and Africa's Great Lakes region.

There are 24 countries in the region that are wholly or partly covered in this guidebook (25 if one includes the territory known as Western Sahara that is claimed by Morocco as its own). Most are desert lands, yet despite this they have a rich and diverse mammal fauna. We have been lucky to have seen and worked with many of these species in the wild: we rediscovered the Arabian Tahr in the United Arab Emirates (UAE) long after it was believed to have become extinct in that country. We were also the first to discover Blanford's Fox occupying the hill country of the UAE and northern Oman, the Egyptian Spiny Mouse in the UAE and the Nile Grass Rat occupying the monsoon woodland of the Omani Dhofar. There were other discoveries, some expected and others not.

Our discoveries resulted from the fact that many areas in the region are poorly known zoologically. Due to warfare and other factors, some countries do not lend themselves to zoological exploration, and a number have been no-go areas for decades. Much of our information comes from the reports of explorer naturalists dating back 50 years and more. Despite this, we have tapped a wealth of information from filing cabinets filled with unpublished reports, our own fieldwork, papers in scientific journals, the shelves and drawers of museums, and in some cases by word of mouth.

More than 4,800 species of mammal are recognized worldwide, of which about 332 are known from North Africa and the Middle East. We believe not only that with further exploration more species will come to light, but also that improved taxonomic technologies will reveal some species to comprise a complex of species.

The region's total includes 29 marine mammals, including 2 seals – the endangered Mediterranean Monk Seal (*Monachus monachus*) and the Caspian Seal (*Phoca caspica*), which lives only in the Caspian Sea – and the Dugong (*Dugong dugon*), which in this region is known only from the Persian Gulf and the Red Sea. The remaining 26 marine mammals are members of the family Cetacea, the whales and dolphins. The marine mammals are generally poorly known in the coastal waters of the region, and it is certain that more species will be discovered over time. Because of the difficulty of observing these marine creatures in the wild state, we decided reluctantly to omit them from this book.

Mammal watching can be tricky. Many of the smaller species – bats, shrews and small rodents – are nocturnal, secretive and rarely seen. Yet from time to time you will encounter some of these fascinating creatures. All the larger mammals occurring in the region are covered here, as well as selected smaller species that are representative and typical of their genus or family. Among the smaller species included are diurnal animals (which are more easily seen than nocturnal ones), and nocturnal species that are distinctive enough to enable you to at least decide to which family your sighting belongs. It is difficult to identify the species of, for example, a jerboa you might see at night, but armed with this guide you will at least be able to pin it down to the group of 'mini-kangaroo-like' rodents. Some rodent groups lend themselves to this kind of coverage, but others are nondescript, obscure and limited in their distribution – these have not been included, although we have presented a few examples of such groups to at least let you identify broad groupings, such as horseshoe and free-tailed bats in the bats group.

HOW TO USE THIS BOOK

This book has been compiled to enable easy identification of the larger and more conspicuous mammals of the countries in the region. A few smaller species have also been included. Photographs illustrate differences between the sexes where necessary, and portray young where they differ significantly from adults.

Measurements used for the species are: total length (nose tip to tail tip) for smaller species, shoulder height for larger mammals, wingspan for bats and weight. Identification, habitat and behaviour, and distribution and status are included under separate headings in all major species descriptions. An indication of where a species is known to occur in conservation areas is provided at the end of most accounts. Here acronyms have been used as follows: AOS, Arabian Oryx Sanctuary; CA, Conservation Area; FR, Forest Reserve; PA, Protected Area; NP, National Park; NR, Nature Reserve or National Reserve; NSF, National State Forest; R, Reserve; SF, State Forest; WR, Wildlife Reserve.

The vast majority of the parks, reserves and refuges in the region have never been fully surveyed, so in many cases only the larger or more obvious species are known to occur in these areas. Where relevant, the accounts also include brief descriptions of similar species whose status in the region is uncertain.

HABITATS

This account of the principal habitats across the region is a simplified version of the real situation. Each area comprises many different vegetation types, but such detail is unnecessarily complex for a book of this scope. Many mammal species are restricted to a particular vegetation type or habitat, while others range over several types. Some are specialists and still others generalists.

Desert

North Africa is dominated by the Sahara, a mixture of sand ergs, gravel plains and vast rocky massifs that rise high above the desert floor. Much of Arabia is desert land – including the great sand sea of Rub al'Khali – that comprises a mixture of sand, gravel and rocky outcrops and ranges. A considerable area of the great central and eastern plateaux

of Iran is classified as desert, including the salt desert known as Dasht-e Kavir. True desert is characterized by very low annual rainfall (less than 100 mm). There is often little or no rain for long periods. As a result vegetation is very sparse and in some areas there may be none at all for several years at a time.

Semi-desert

Semi-desert is not defined as clearly as true desert, but includes the Sahel belt to the south of the Sahara and large areas in the Middle East. Annual rainfall is unpredictable: less than 500 mm (often considerably so), but not usually below 100 mm. Vegetation is more apparent than in desert areas and may include some trees and bushes, often

with strong seasonal growth of annual and perennial grasses and herbaceous plants following rain. Wherever semi-desert is found, vast tracts have often been severely degraded by livestock overgrazing.

Montane Areas

These areas include the Atlas ranges of north-west Africa, the Saharan massifs, the ranges of western Saudi Arabia and Yemen, those of northern Oman, the Zagros and its extension eastwards, the Makran range, Alborz of Iran and the numerous systems that rise up over much of Turkey. Vegetation and rainfall vary considerably between ranges

and at different altitudes. Many ranges attract differing amounts of rainfall on one face or the other: one side may receive higher rainfall and harbour good vegetation cover; the other may be dry and sparsely vegetated. A number of ranges receive heavy snowfall in winter. Many of these ranges have also been heavily grazed and browsed by domestic livestock.

Forest

The once-great forests of the region have been greatly depleted and they are associated with the better-watered mountainous areas. As is the case with so many habitats in the region, destruction and modification of forests has been considerable. A depressing example of this is the once-mighty forests of the Caspian hill and mountain slopes, where only

about 8 per cent of what was present at the turn of the 20th century still remains. The structure of forests varies depending on the ranges and even on different parts of the same range. In the Atlas ranges there are cedars, pines, juniper, oaks and thuya. The Zagros has remnant oak forests. The Alborz hosts a mixture of deciduous trees on its north-facing slopes, while oak forest dominates in its upper reaches and remnant patches of juniper grow on its southern slopes. The Iranian Caspian coastal plain region has lost all its lowland forests. Turkey's forests are strongly associated with mountain ranges and contain a mixture of conifers and deciduous species. The most densely forested area in Turkey is situated along the Black Sea coast.

Mediterranean Evergreen Scrub

This vegetation type is found on the coastal plain and north-facing slopes of the Atlas in north-west Africa, from Morocco across the top of Algeria to northern Tunisia. It is also present on the eastern coast of the Mediterranean Sea, across the south, west and north-west of Turkey, and to a lesser extent in northern Syria and Iraq. The general term for this

vegetation type is maquis. It consists mostly of tall, dense evergreen shrubs and low trees, but varies greatly in species composition from area to area. Various species of oak are associated with certain maquis systems. The countries in the region with the greatest extent of maquis vegetation are Turkey, Morocco and Algeria. Exploitation of forests in this vegetation type has been going on for thousands of years and they are but a shadow of what they once were. Overgrazing and clearing of maquis is an ongoing problem.

SUCCESSFUL MAMMAL WATCHING

There are a number of basic steps to follow in order to identify a mammal you have seen. Initially try to establish the group to which the mammal belongs. For example, is it an antelope, dog or cat? Then try to make an estimate of shoulder height or, in the case of smaller species, total length and tail length: is the tail shorter or longer than the head and body length? Are any features very obvious and thus distinguishing: a bushy or sparsely haired tail; spots or stripes; particularly long or short ears? Make a note of the animal's general colour and any distinct markings.

Also make a note of behavioural traits or peculiarities. Was the animal in a large group or alone? Was it in a tree? Did it enter a burrow? Once you have sufficient information you can consult this guide and try to establish the species.

A useful indicator of an animal's presence in an area is its spoor, or tracks – most easily seen in sand or mud – or its droppings. Fresh tracks might give you a clue as to what the animal was doing and might even lead to a glimpse of a more secretive species.

CONSERVATION

The actions of humans in North Africa and the Middle East, as in most other parts of the world, have resulted in extinctions, declines in populations, and destruction and modification of habitats. During the colonial era hunting on a large scale began. The Second World War left a legacy of an abundance of weapons, and versatile Jeeps became available: the result was a considerable slaughter of game animals. Even before this, hunting was taking its toll. Consider the princes and grandees of Persia (now Iran), in the late 19th century. One case is sufficient to explain the scale of the slaughter. At the age of 14, a Qajar prince, Zell-e Soltan, and his entourage went on a hunting expedition to the Mian Kaleh Peninsula, which extends into the Caspian Sea. The peninsula covers over 60,000 ha and was rich in game species. Among their bag were 150 Red Deer stags, 18 Leopards and 35 Tigers.

The Tiger is now extinct in the region, but just over 100 years ago it roamed from eastern Turkey and across all of northern Iran. This distinct race, the Caspian Tiger (*Panthera tigris virgata*), disappeared from Turkey but continued to survive in Iran until, it was claimed, the last of these magnificent cats was shot in 1953 in what is today the Golestan National Park to the east of the Caspian Sea. In 1958 there was a reliable sighting of a Tiger in this same area.

Although hunting played some role in the demise of the Tiger and the Asiatic Lion (*Panthera leo persica*), it came mostly from the loss of their forest and woodland habitat and the decline of their prey animals. Lions lived in North Africa and parts of the Middle East up to the middle of the 19th century. They were wiped out in Palestine during the Crusades, but survived in eastern Turkey until about 1870. In Iran they occurred across the south, from the Zagros Mountains to Baluchestan, with the last reliable sighting made in 1942 in the south-west 65 km from the town of Dezful. Lions still occurred in the great marshes at the confluence of the Euphrates and Tigris Rivers, Iraq, in the mid-1940s. The only surviving Asiatic Lions now live outside the region in the Gir Forest of western India. In North Africa, lions had gone from the Moroccan coast by the mid-1800s, although it is believed that some survived in the high Atlas until the mid-1940s.

It was not only the big cats that suffered. At least seven species of antelope were driven to total or regional extinction. The Red Gazelle (*Gazella rufina*) of North Africa (Algeria) is known from just three museum specimens collected in the 19th century. The Arabian Gazelle (*Gazella arabica*) is known from a single museum specimen said to have been collected on the Farasan Islands (the remaining gazelles on these islands are not of this species). The Saudi Gazelle (*Gazella saudiya*) is extinct in the wild, and the few held in captivity are now believed to be possible hybrids. The Bubal Hartebeest (*Alcelaphus b. buselaphus*) that once lived across large swathes of North Africa had probably disappeared from the region by the 1930s. The extinction of the Queen of Sheba's Gazelle (*Gazella bilkis*) from its limited range in Yemen is almost certain, but one can always hold a little hope for its survival.

Then there was the demise of the region's two oryx: Arabian Oryx (*Oryx leucoryx*) once roamed the Arabian Peninsula, and Scimitar-horned Oryx (*Oryx dammah*) lived in the Sahara and Sahel. Some of these species were once abundant. In 1936 a herd of Scimitar-horned Oryx estimated to be 10,000 strong was seen in north-central Chad. By the 1980s a few hundred were believed to survive in that country; today the species is extinct as a truly wild mammal. Its fate was sealed when Libyan-backed rebels and government forces fought for dominance in the area from 1978. Fortunately, populations of both oryx were held in captivity and these can now be used in reintroduction programmes.

Other species are hovering on the edge of the abyss with little being done to save them. One other possible extinction may have taken place in Arabia in the 1960s, although details remain sketchy. The Lesser Kudu (*Tragelaphus imberbis*), a well-known antelope in East Africa, is known in the region from just two individuals that were shot in Arabia – in Saudi Arabia and Yemen – where it lived in mountainous country associated with the wadi systems and adjacent plains. Local tribesmen say it does not occur in the region today.

Even up to the time of the Islamic revolution in 1979, substantial numbers of game survived in Iran. In the Golestan National Park in that year there were said to be as many as 15,000 Urial (Wild Sheep) and large numbers of gazelles across the country. Following the revolution, it is believed that as much as 80 per cent of Iran's larger wildlife was lost.

The current state of conservation across North Africa and the Middle East is a mixed bag. Some countries have minimal conservation legislation and no reserve management. These contrast with a few that are making major efforts to protect their habitats and species. On paper, the conserved areas are substantial, but on-the-ground protection is often minimal to non-existent. Human populations continue to grow, and their livestock continues to overgraze and trample, out-competing the wild ungulates. In those areas where it is practical to cultivate, every square centimetre is made use of. There is also continuing armed conflict in parts of the region. There are conflicts in Iraq, the Western Sahara (Morocco), Israel and Palestine, and Lebanon, and conflict with the Kurds in eastern Turkey. We have little idea of what is happening to the wildlife in many of these areas.

If core wildlife populations are protected they can bounce back, but once an animal is driven to extinction only a few skins and skulls will remain in museum collections.

North A

N

Mediterranean

Algiers ■ Tunis ■
TUNISIA
Rabat ■ ●31 ●24
●30 ●29 ●25 Tripoli ■
MOROCCO Atlas Mts. ●25 ●26 ●22
●32 ALGERIA ●23 L

A T L A N T I C
MAURITANIA S A H A R
●28
●27

●33 MALI ●34 NIGER CH
Nouakchott ■ NIGER

O C E A N *R. Niger* Niamey ■
■ Bamako N'Djamena ■

Key

--- National boundaries
● National parks and reserves

IRAN
1 Golestan
2 Touran
3 Kavir
4 Khabr
5 Urumiyeh

TURKEY
6 Olimpos-Beydaglari
7 Köprülü Kanyon
8 Munzur
9 Semdinli-Rubaruh

SAUDI ARABIA
10 Harrat al-Harrah
11 Hawtat Bani Tamim
12 Mahazat as-Sayd
13 Raydah Escarpment

OMAN
14 Jiddat al Harasis
   (Oryx reserve Yaluni)
15 Jabal Samhan

JORDAN
16 Zubiya
17 Azraq Oasis

ISRAEL/PALESTINE
18 S. Arava Valley

Middle East National Parks and Major Reserves

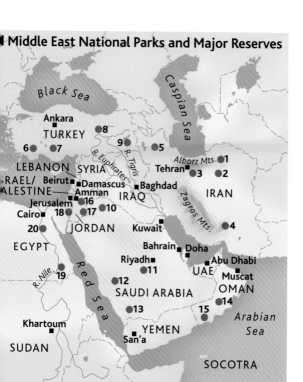

EGYPT
19 Wadi el Allaqui
20 Wadi el Raiyan

LIBYA
21 El Kouf
22 Tripoli
23 Zellaf

TUNISIA
24 Djebel Chambi
25 Bou-Hedma
26 Sidi Toui

ALGERIA
27 Ahaggar
28 Tassili
29 Saharan Atlas

MOROCCO
30 Toubkal
31 Tazzeka
32 Souss-Massa

MAURITANIA
33 Banc d'Arguin

NIGER
34 Aïr & Tenere

CHAD
35 Quadi Rime – Ouadi Achim

SPECIES DESCRIPTIONS

EVEN-TOED UNGULATES (Ruminantia)

This order includes all antelopes, sheep, goats and deer in the region. Like all ruminants, they chew the cud and have incisor teeth only on the lower jaw. In the region, males of all species carry horns, or antlers in the case of deer, as do females of many species.

ANTELOPES

ARABIAN ORYX *Oryx leucoryx*

Identification This is the largest antelope on the Arabian Peninsula. It is powerfully built and both sexes carry long, rapier-like horns (60–70 cm). The overall white appearance and size distinguishes it from other species. The lower legs are dark brown to black with a white ring above each hoof. The tail is of moderate length and white for most of its length, but terminating in a tuft of black hair. The face has a black partial blaze, and a black patch between the horns and on the upper throat. Calves are grey at birth, but soon take on a pale brown colour. At three months they start to take on adult coloration. The large, rounded hoofs of this antelope are an adaptation to its desert habitat.

Size Shoulder height 90 cm (to 1.2 m).
Weight To 120 kg.

Habitat and Behaviour This is a species of sand and stony desert plains, as well as sand dunes. It avoids mountainous country. It does not need to drink, obtaining sufficient moisture from the plants that it eats. Although they are mostly grazers, Arabian Oryx browse readily during the driest periods. They are nomadic and move considerable distances in search of fresh grazing grounds. During the hottest months they feed mainly at night. Herd size is generally small, ranging from 2 to 30 individuals with a dominant bull and a dominant cow leading the herd. Usually a single calf is dropped after a gestation period of about 260 days.

Distribution and Status Once common across the Arabian Peninsula, the Arabian Oryx occurred as far north as the Euphrates River and to the Mediterranean coast of Palestine. By 1972 it had been hunted to extinction in the wild, but there were captive populations from which reintroductions into the wild were made in Wadi Yaluni, Oman, in the 1980s. Since then populations have been reintroduced in conservation areas in Saudi Arabia and Jordan. Privately held captive herds are maintained in a number of Arabian countries. Substantial herds are held outside the region.

Conservation Areas Mahazat as-Sayd PA, Uruq Bani Ma'arid PA (Saudi Arabia); Yaluni AOS (Oman); Shaumari WR (Jordan); Hai Bar NR (Israel).

Arabian Oryx cows and uniformly coloured calf

SCIMITAR-HORNED ORYX *Oryx dammah*

Identification This is a large, powerfully built antelope. Both sexes carry long horns (>1 m) that sweep backwards in a curve. The overall body coloration is white to dirty white, with brown to chestnut on the neck and chest. There are paler brown markings on the rump and upper back legs. The face is white with a brown blaze. The tail is well haired and mainly white, but darker towards the tip. Calves are fawn in colour.

Size Shoulder height 1.2 m.
Weight 200 kg.

Habitat and Behaviour This is a species of semi-desert country that penetrates true desert in search of fresh grazing. Although it is mainly a grazer, it will readily take browse if grasses and herbs are not

13

Long, swept-back horns are a characteristic of this species

available. Scimitar-horned Oryx once lived in herds of 20–40 animals, and several hundred animals would gather in areas where rain had fallen. They were subject to frequent nomadic movements in search of grazing. A single 15 kg calf is born after a 242–56-day gestation period.

Distribution and Status This species once occurred across much of the Sahel and penetrated deep into the Sahara. It may now be extinct in the truly wild state, but remnant populations may survive in northern Niger and Mali. Large captive populations are held and are being used in reintroduction programmes.

Conservation Areas Souss-Massa NP (Morocco); Bou Hedma NP (Tunisia); Tripoli R (Libya).

ADDAX *Addax nasomaculatus*

Addax cow – both sexes carry long, spiralled horns

Identification The Addax is a fairly stocky antelope with large, rounded hoofs. Both sexes carry long, diverging and spiralled horns (80 cm), and have a mat of coarse, dark brown hair on the forehead. The overall body colour is smoky-grey, with paler to white rump, underparts and legs. A distinct white chevron crosses the face between the eyes.

Size Shoulder height 1 m.
Weight 80–120 kg.

Habitat and Behaviour The Addax is a species of sand dunes and gravel plains in the Sahel and Sahara. In the past this antelope migrated in herds numbering hundreds, seeking new plant growth following rare rain showers. Like most other desert antelopes, Addax graze and browse, and are independent of drinking water. The average size of a herd is around 20 animals, and the home range of the antelopes can extend over several thousand square kilometres. The single calf of up to 7 kg is born after about a 260-day gestation period.

Distribution and Status Like other desert-dwelling antelopes, Addax have been hunted ruthlessly. Perhaps as few as 250 individuals survive in the wild. Some of these may range in southern Algeria and Libya, but most survivors are in the Sahelian countries of Mauritania, Mali, Niger and Chad. Substantial numbers are held in captivity, and some are being used in reintroduction programmes.

Conservation Areas Souss-Massa NP (Morocco); Ahaggar NP (Algeria); Bou-Hedna NP (Tunisia); Tripoli R (Libya).

DAMA GAZELLE *Gazella dama*

Identification This is the largest of the true gazelles, with proportionately long legs and neck, and slender build. Both sexes carry horns (33 cm), which are short, bent back strongly from the base, curved at the tips and strongly ringed. There are three colour forms: the western *G. d. mhorr* has the greatest extent of rufous hair over the body; the eastern *G. d. ruficollis* (Red-necked Gazelle) is predominantly white with rufous only on the neck and shoulders; the central *G. d. permista* is intermediate in the extent of white and rufous. The underparts and rump are bright white in the three races, all of which also have a distinct white throat spot.

Size Shoulder height 90–120 cm.
Weight 40–75 kg.

Habitat and Behaviour This is a gazelle of the desert and its fringes. At one time it occurred right across the Sahel belt and penetrated the Sahara both from the south and north. It used to form into large, temporary nomadic herds searching for fresh plant growth. These movements followed a set pattern with a southwards movement during the driest months, which turned northwards when the sparse rains fell in the Sahara. Much of the time the gazelles lived in small herds of females and their young, rarely of up to 15 individuals, accompanied by a single adult ram. These herds mingled freely with the more abundant Dorcas Gazelle. Dama Gazelle are mainly browsers but they readily take grass and herbaceous plants. Birthing is synchronized with the onset of the meagre rains – the single fawn is dropped after a gestation period of about 198 days.

Distribution and Status The once extensive range of the Dama Gazelle stretched from the Atlantic coast to the Sudan, penetrating northwards from Morocco to Libya and possibly as far east as western Egypt. Numbers of this species – like other gazelle species – have been greatly reduced by hunting, competition with domestic stock for food and prolonged drought. Probably no more than 500 Dama Gazelle survive, with the last North African animals believed to inhabit the south of Algeria. Widely scattered, remnant

Western race of the Dama Gazelle

Eastern race, also called the Red-necked Gazelle

populations may continue to exist in Mauritania, Mali, Niger and Chad. Reintroductions of Dama Gazelle have been undertaken in two countries.

Conservation Areas Souss-Massa NP (Morocco); Ahaggar NP (Algeria); Bou-Hedma NP (Tunisia).

SLENDER-HORNED GAZELLE *Gazella leptoceros*

Identification Known also as the Sand, Rhim or Loder's Gazelle, the Slender-horned Gazelle is the palest of all North African gazelles. It has a dull greyish-fawn coat, white underparts and indistinct facial markings. The distal half of the short tail is black. The horns (35 cm) are slender, especially those of the female, and well ringed. The rams have almost straight horns.

Size Shoulder height 65–70 cm.
Weight 20–30 kg.

Habitat and Behaviour The Slender-horned Gazelle is an antelope of mainly sandy desert that extends its range into adjacent hill country. A mixed feeder, both grazing and browsing, like most desert antelopes this species is independent of drinking water. It is strongly nomadic and lives in family groups or small herds, although larger temporary groupings may come together at productive feeding grounds. Each small group of females and associated young is accompanied by an adult ram.

Distribution and Status It is estimated that less than 5,000 individuals survive over the vast Saharan range of the Slender-horned Gazelle; it has disappeared from much of its former range. Populations are known from north-central and south-east Algeria, southern Tunisia, Libya and north-west Egypt. Small populations are said to persist in north-central Niger.

Conservation Areas Tassili NP (Algeria); Djebil NP (Tunisia); Zellaf NR (Libya); Aïr and Tenere NR (Niger).

CUVIER'S GAZELLE *Gazella cuvieri*

Identification Known locally as the Edmi, this gazelle has dark grey-brown upperparts and clean white underparts separated by a broad, dark lateral band, and distinctive dark and pale facial markings. It is the only gazelle likely to be encountered in the Atlas region of north-west Africa, although it may overlap with the much paler Dorcas Gazelle in some of its range. Its horns are heavily ringed and of moderate length (30 cm); they are shorter and more slender in ewes.

Size Shoulder height 60–80 cm.
Weight 35 kg.

Habitat and Behaviour Cuvier's Gazelle exists in a range of habitats from open forest to rocky desert in the Atlas ranges and adjacent areas. It is a mixed feeder that grazes and takes browse. It never forms into large herds, living in small herds usually numbering 3–5 individuals. Each herd is generally controlled by a territorial ram within a relatively small home range. The species is not known to be

Right: Slender-horned Gazelle rams have almost straight horns

Cuvier's Gazelle, showing its very distinctive markings

nomadic, unlike many North African antelopes. Surprisingly little is known about it, but it is believed that births are seasonal and twins are not unusual.

Distribution and Status Cuvier's Gazelle is largely restricted to high country in Morocco and northern Algeria, with a small population in western Tunisia. Populations have been decimated by hunting and competition with domestic livestock; only about 1,000 individuals survive.

Conservation Areas Saharan Atlas NP, Belezma NP, Mergueb State NR, Djebel Senalba NSF (Algeria); Djebil Chambi NP (Tunisia).

DORCAS GAZELLE *Gazella dorcas*

Identification This is a small gazelle that lacks distinctive features. The overall body colour is pale- to sandy-fawn, with white underparts, inner legs and buttocks. The lateral body stripe is usually pale and indistinct, especially when observed from a distance. The short black tail contrasts with the white inner buttock region. There are white stripes on the face and under the jaw, and a distinctive brown blaze on the face. The horns are not particularly long (25–37 cm), lyrate (lyre-shaped) and strongly ringed; the horns of females are more slender and shorter than those of males.

Dorcas Gazelle ram

Size Shoulder height 60 cm.
Weight 15–>20 kg.

Habitat and Behaviour The Dorcas Gazelle shows a preference for desert and semi-desert plains, extending onto sandy areas. It occurs across much of the Sahara and its fringes, but is absent from the coastal plains except along the Red Sea. Like most desert antelopes, Dorcas Gazelle are mixed feeders, taking grasses, shrubs and browse. They are usually seen in herds of 20 or more, but larger groups may be encountered moving to areas with new plant growth. In areas of true desert, herds are smaller and pairs and solitary animals are frequently encountered. Non-territorial rams often form into bachelor groups. Usually a single lamb weighing 1.5 kg is born after about a 164-day gestation.

Distribution and Status This gazelle is present in all North African countries – including those in the Sahel zone from Mauritania to Sudan – and extends into the Egyptian Sinai and Israel and Palestine. Numbers have been greatly reduced by hunting. It is estimated that more than 10,000 animals survive across its vast range.

Conservation Areas Souss-Massa NP (Morocco); Tassili NP, Ahaggar NP (Algeria); Bou Hedma NP, Sidi Toui NP, Djebil Chambi NP (Tunisia); New Hisha NR, El Kouf NP (Libya); Aïr and Tenere NR (Niger); Ouadi Rime-Ouadi Achim FR (Chad); Mezukai Hazinim NR, Har Hanegev NR, Hanahalim Hagdolim NR, Masiv Elat NR (Israel).

MOUNTAIN GAZELLE *Gazella gazella*

Identification The Mountain Gazelle is also known as the Idmi, or Arabian Gazelle. It is a small, slender and elegant gazelle with short, lyrate horns (25 cm) that are strongly ringed; those of ewes are more slender and shorter than those of rams. Several races of Mountain Gazelle are recognized that vary in colour, but all are darker coloured above and bright white below. The neck, chest and much of the legs are a somewhat lighter brown. A lateral line of light to darker brown separates the upper and lower parts. On the face there are white stripes from the eyes to the muzzle, as well as a brown blaze. The tail is short and black-haired.

Size Shoulder height 60 cm.
Weight 12–20 kg.

Habitat and Behaviour This gazelle has a very wide habitat tolerance and occupies coastal plains, sand and stony desert, hill and mountain country, and wadi bottoms. Much of its range seems to be closely associated with acacia trees and bushes. It is most often seen singly or in small groups, with a ram controlling a territory and the ewes within it. Males mark their territory with piles of dung pellets and secretions from glands in front of the eyes, which they rub on twigs and grass stalks. The gazelles are mostly active in the day but may feed at night, taking browse and graze equally. The young may be born at any time of the year, but seasonal peaks occur in some areas of the gazelle's range.

Mountain Gazelle ram – note the white facial stripes

22

Newborn Mountain Gazelle fawn

Distribution and Status The Mountain Gazelle occurs in the UAE, where only the Jebel Ali sand desert holds a substantial population, with populations scattered along the coastline of Oman. Yemen and western Saudi Arabia populations have been hugely reduced by hunting. The largest populations are located in Oman, Israel and Palestine.

Conservation Areas Ibex Reserve, Uruq Bani Ma'arid PA, Jebel al-Rar, Majami al-Hadb PA (Saudi Arabia); Wadi Sareen Tahr R, Yaluni AOS (Oman); Wadi Mujib WR, Wadi Dana-Finan WR (Jordan); Ya'ar Yehudia NR, Mezukai Herev NR, En Gedi NR (Israel).

SAND GAZELLE *Gazella subgutturosa*

Identification This fairly stocky gazelle is also known as the Rheem, or Rhim, in Arabia, and the Goitered Gazelle in its eastern range, including Iran. It is light fawn-brown above and white below, as well as on the rump and face. Animals in some populations are very pale and may look almost white from a distance. The gazelle has a distinctive short black tail, which is held erect when running. The ears are noticeably long and pointed. The horns are fairly long (34 cm), lyrate, strongly ringed and very close together at the bases; those of ewes in western populations (*G. s. marica*) are more slender. Ewes in eastern populations (*G. s. subgutturosa*) either do not carry horns, or have very short horns. A swelling on the upper throat, resembling a goitre, gives these animals their scientific name, and the common name of eastern populations. The eastern subspecies averages larger than that of the west.

Sand Gazelle ewe of the western population

Sand Gazelle ram has heavier horns than female

Size Shoulder height 75 cm.
Weight 22–42 kg.

Habitat and Behaviour This is a gazelle of sand dunes, sand and gravel plains, and upland plateaus, but not mountain country. It penetrates deeply into sand seas such as the Rub al'Khali in southern Saudi Arabia. Within its Iranian range it mainly occupies semi-desert and desert fringes. It lives in small herds numbering 2–10 animals; larger groups may gather temporarily at productive feeding grounds. Unlike other gazelle species, Sand Gazelles run in tight formation with neck outstretched and short black tail held erect. They eat small shrubs and grass, and apparently rarely browse. Most lambs are dropped in the cooler months in March–May over much of the gazelle's range. Usually a single lamb is born, but twins are not uncommon.

Distribution and Status Numbers of Sand Gazelle have been greatly reduced right across its range due to hunting and competition for grazing with domestic stock. It once ranged across the Arabian Peninsula, eastwards through Syria and Iraq, into south-east Turkey and throughout Iran where suitable habitat existed. Today it is restricted largely to conservation areas and the most isolated regions, with probably less than 10,000 animals surviving in the wild. The largest (although greatly reduced) populations survive in Iran.

Conservation Areas Mahazat as-Sayd PA, Uruq Bani Ma'arid PA (Saudi Arabia); Shaumari WR (Jordan); Ceylanpinar SF (Turkey); Bamou NP, Golestan NP, Kavir NP, Sorkeh Hesar NP (Iran).

CHINKARA *Gazella bennettii*

Identification The Chinkara, Jebeer, or Indian or Bennett's Gazelle is represented by at least two subspecies in Iran. The northern and western form (*G.b. shikarii*) is very pale and could be confused with the Sand Gazelle. The darker form (*G. b. fuscifrons*) occurs in the south-east and along the Makran coast to the border with Pakistan. These animals are light to darker brownish-yellow above and white below and on the inner buttocks, and have a variably brown blaze on the face. The lateral line separating the upper and lower body is indistinct or slightly darker than the dorsal hair. The tail is short and dark in colour, and the ears are long and pointed. The horns are very similar to those of the Sand Gazelle, reaching 35 cm in the male. Some authorities believe that the Chinkara belongs in the Dorcas group.

Size Shoulder height 55–65 cm.
Weight 15–25 kg.

Habitat and Behaviour The Chinkara occupies mainly semi-arid areas, particularly desert fringes where there are gravel and scree plains, or hilly terrain. It is usually seen singly, or in small herds of up to six animals. Chinkara are mixed feeders that graze where available but also browse. They seek out succulent plants for their water content, but like most arid-dwelling animals they are water independent. Lambs have been recorded in May in south-west Iran, but also in September, perhaps suggesting a double birthing. In the north they are born in late autumn.

Distribution and Status The Chinkara only occurs to the east of the Zagros Mountains in Iran. Remaining populations are concentrated in the central and south-eastern parts of Iran, including the coastal plain of the Gulf of Oman. It now survives mainly in conservation areas, and is known from nine of these.

Conservation Areas Kavir NP, Khabr-va-Rouchoon WR, Bahram-e-Gour PA (Iran).

Chinkara ram

Chinkara ewe – note the black tail

Similar Species The Queen of Sheba's Gazelle (*Gazella bilkis*) deserves mention here, even though it may well be extinct. It superficially resembled the Mountain Gazelle (*G. gazella*), but was probably shorter in the leg and had much straighter horns. The main character was the very dark colour of the coat that contrasted strongly with the white underparts. It was known to occur only on the euphorbia-vegetated hill slopes in the vicinity of Ta'izz in south-western Yemen. It is discussed here because there is a slim chance that individuals may survive in this zoologically poorly known country.

RED DEER *Cervus elaphus*

Identification The population of Red Deer in Algeria/Tunisia is known as Barbary Red Deer, while the Red Deer in Iran are called Maral. A number of subspecies are described across the region, but all are easily recognizable as Red Deer. This is by far the largest deer in the region. Only the stag carries the large, multi-branched antlers that are shed and regrown each year. Stags in Iranian populations usually have a double brow-tine; elsewhere this is usually single. The winter coat is grey-brown to dark-brown, with or without faint white spotting (if present, spotting is clearer in the summer coat). The summer coat is usually more yellowish- to reddish-brown. A dark dorsal stripe is generally present and runs from the neck to about two-thirds the length of the back. There is also usually a pale yellowish rump patch. The underparts are paler than the rest of the body.

Size Shoulder height 75–110 cm.
Weight 75–>200 kg.

Habitat and Behaviour Red Deer have a relatively wide habitat tolerance, but are typically found in forest and open bush cover, often (but not always) in hill country. They eat a wide range of plants that they graze and browse equally. For most of the year stags and hinds live in separate herds, with each sex normally forming groups of 8–20 individuals. Herds of hinds and young are usually quite stable, but those of stags are much more loosely associated. Older Red Deer stags are often solitary. During the rut (September to October in North Africa; late summer to autumn in Iran), a mature stag moves into an area occupied by a herd of hinds. The dominant hind continues to control the herd, with the stag trying to keep other males away. It is

Barbary Red Deer hind in Algeria

Red Deer stags with antlers 'in velvet'

at this time that stags give their impressive roaring displays, which are intended to intimidate other stags (and to attract hinds). Should these displays and posturing fail, stags may fight, which sometimes results in severe injury and even death. A single fawn (very rarely twins) is dropped after a gestation of about 235 days. The fawn remains hidden for 2–3 weeks after birth.

Distribution and Status The naturally occurring North African population is restricted to extreme north-eastern Algeria and the adjacent area of Tunisia, with about 4,000 individuals. Animals in Morocco were introduced from Spain. The Red Deer is absent from Arabia. Scattered populations (perhaps 5,000 animals) occur in pockets of suitable habitat in Turkey. In Iran heavy hunting pressure and habitat degradation have caused the extinction of the species over much of its former range. It is known only to survive in forests of the Alborz ranges towards the Caspian Sea. Numbers are unknown.

Conservation Areas El Kala NP, Bouchegouf NR (Algeria); El Feidja NP (Tunisia); Dodangeh WR, Golestan NP (Iran).

PERSIAN FALLOW DEER *Dama mesopotamica*

Identification Sometimes called the Mesopotamian Fallow Deer, this species is the second largest and rarest deer in the region. The summer coat is overall light brown, but in winter it is darker brown and washed with grey. The lower neck, chest and belly are much lighter in colour and often white. The back, upper flanks and thighs have numerous white spots, and a white lateral line is often present along each side. The well-haired tail is predominantly white. Only stags carry antlers, and these are shed annually in February and fully regrown by July. The antlers are considerably shorter than those of Red Deer.

Size Shoulder height 85–110 cm.
Weight 70–180 kg.

Habitat and Behaviour Persian Fallow Deer once occurred in a wide range of forest and woodland types, but now only survive in dense riverine forest and thicket. They are mainly browsers, feeding from a number of trees (especially poplars and tamarisk), and will also graze, and eat moss and fungi. Little is known about their behaviour, but the hinds and young form into small herds. The stags live alone or in small

29

Persian Fallow Deer stag

Persian Fallow Deer hind

bachelor groups outside the breeding season. The rut takes place in September, and dominant stags acquire harems. The fawns are born between mid-March and mid-April, and remain hidden for the first two weeks of life.

Distribution and Status This is the Asian form of Fallow Deer. It once occurred eastwards from the Mediterranean coast through Syria, Iraq and southern Turkey, and across western Iran. By the 1950s it was thought to be extinct, but in 1957 it was found to have survived in small numbers in forest along the Karkheh and Dez Rivers in Iranian Khuzestan, close to the border with Iraq. Breeding programmes have resulted in translocations to several Iranian protected areas. The

allow Deer (*D. dama*) once occurred along the Mediterranean coast of
frica, but it is now extinct. A breeding and reintroduction programme
was started in 1966 for this species at Antalya in south-west Turkey.
he Persian Fallow Deer is sometimes considered a subspecies of the
allow Deer.

Conservation Areas Dasht-e Naz WR, Karkheh WR, Dez WR, Lake
Jrumiyeh (Ashk Is.) (Iran); Hai Bar NR (Israel).

ROE DEER *Capreolus capreolus*

Identification The smallest of the three deer species in the region,
he Roe Deer has a largely uniform red-brown to yellow-brown coat,
white rump patch and very short tail (2–3 cm). The winter coat usually
has a greyish tinge. Fawns have heavily white-spotted coats. Only the
ams carry the short (20–30 cm) antlers, with usually 3–6 points. The
ntlers are shed in mid-autumn and regrowth is complete by spring.

Size Shoulder height 70 cm.
Weight 15–30 kg.

Habitat and Behaviour Roe Deer frequent woodland with under-
growth, and heath land. They are mixed feeders that graze and
browse, and eat fallen fruits, fungi, lichen and moss. They are usually
olitary except when the ewe is accompanied by a fawn, but a ram

Only ram Roe Deer carry antlers

may be with them. In areas of suitable habitat they may reach quite high densities and several animals may be in sight when feeding which is usually at its peak in the cooler morning and afternoon hours during summer. During the spring and summer, both sexes defend a territory against members of the same sex. Mating takes place in summer, but ewes can store the fertilized egg for about four months and births occur in spring. This is the only ungulate with this ability. The ewes frequently give birth to twins, but also to single fawns and rarely three. The fawns weigh less than 2 kg at birth.

Distribution and Status Roe Deer once lived in the wooded hills along the eastern Mediterranean, but are probably extinct there now. They have been reintroduced to one reserve in Jordan. They still occur in scattered populations in Turkey, with perhaps some surviving in northern Syria and Iraqi Kurdistan. In Iran they survive in the forests of the Alborz Range along the Caspian, and possibly in the extreme north-west borderlands.

Conservation Areas Several in the Alborz Range (Iran); Zubiya W (Jordan).

GOATS

CHAMOIS *Rupicapra rupicapra*

Identification Only one subspecies of the Chamois – the Anatolian or Turkish Chamois (*R. r. asiatica*) – occurs in the region. This has a somewhat goat-like appearance, with a very short, dark-coloured tail. The summer coat is brownish; the winter coat is darker brown. There are distinctive black bands on the sides of the face, a broad white blaze from between the horns to the nose, and white from the lower jaw on to the throat. Slender black horns (15–20 cm) with strongly hooked tips are carried by both sexes.

Size Shoulder height 75–80 cm.
Weight 30–50 kg.

Habitat and Behaviour The Chamois is a species of high mountains that occupies steep slopes and adjacent forested terrain. It is very agile in broken rock and steep scree-slope areas, and even on uneven ground; it can reach speeds of up to 50 km per hour. It feeds on a variety of herbs, grasses, lichens, mosses and young tree shoots, especially those of pines. The diet may vary from summer to winter. The behaviour of the little-studied Anatolian race of Chamois is probably similar to that of animals living in the European Alps. Ewes and their young usually form into herds of 15–30 individuals. Adult males outside the mating season are mostly solitary. During the autumn rut dominant males drive other rams away. Fights can be vicious and may result in the death of one of the combatants. Lambs weighing 2–3 kg are born in the summer months after a gestation of 170 days.

Chamois have a very restricted range in the region

Distribution and Status Restricted to the mountains of north-eastern Turkey within the region of Anatolia, south of the Black Sea, west from Artvin to just south of Trabazon. Small isolated populations may survive further south and west. The species occurs in ten protected areas and one national park.

Conservation Areas Munzur NP (Turkey).

WILD GOAT *Capra aegagrus*

Identification Believed to be the parent stock of the Domestic Goat (*C. hircus*), there are two subspecies in the region. The Persian Wild Goat (*C. a. aegagrus*) occurs over much of Iran and in Turkey. The Sind Wild Goat (*C. a. blythi*) is only known from north-east Iran adjacent to the Pakistan–Afghanistan border. Males of both subspecies are characterized by a long black beard and very long, scimitar-shaped horns that may reach lengths greater than 1.3 m. Females carry more slender and considerably shorter horns. The body colour is variable, but usually greyish-brown to silvery-brown in males, with females being more yellowish-brown to reddish-grey. Males in the

Wild Goat rams (left) have impressive horns and beards

north-eastern population tend to be more reddish-brown. They have a variable dark dorsal stripe and shoulder band. Both sexes have dark and light markings on the fronts of the legs.

Size Shoulder height 70–100 cm.
Weight 25–95 kg.

Habitat and Behaviour The Wild Goat occupies mountainous terrain, rugged country in the central desert, sea cliffs and deciduous forests in the north of Iran. Goats both graze and browse readily. Herds usually number 5–25 animals, but this varies from area to area and season to season. Females and their young associate in nursery herds. For much of the time males form into small bachelor herds or are solitary. During the autumn mating period, they join the nursery herds and compete for mating rights. After a gestation period of about 160 days, 1–2 kids are born, each weighing about 2 kg.

Distribution and Status The Wild Goat inhabits southern and eastern Turkey, may still be present on the Iraqi side of the Zagros Range and is widespread in Iran. Uncontrolled hunting, habitat destruction and competition for food with domestic goats and sheep has reduced Wild Goat numbers dramatically across almost the entire range in the past 30 years. Any impact of interbreeding with domestic and feral goats – numerous throughout the Wild Goat's range – is unknown and little studied. It occurs in 38 conservation areas in Turkey and at least 50 in Iran.

Conservation Areas Köprülü Kanyon NP, Munzur Valdisi NP, Olimpos-Beydaglari NP, Sipildag NP, Termessos NP (Turkey); Ariz and Bafq PA, Daranjir WR, Khar Turan NP, Abbas Abad PA, Golestan NP, Alborz-Markazy NP (Iran).

NUBIAN IBEX *Capra ibex nubiana*

Identification Some authorities believe that this ibex of north-eastern Africa and Arabia is a distinct species. Others consider it to be a subspecies of the ibex that occurs across Eurasia. It is large and goat-like. Rams have a longish beard under the chin and massive, semi-circular horns (70–140 cm). The horns have transverse knobs along their entire outer length (except at the tips). Ewes have more slender and much shorter horns. The overall body coloration is dark

Nubian Ibex ram (left) and ewe

yellowish- to reddish-brown, with white to creamy-fawn underparts. Black and white markings adorn the forwards-facing surfaces of the front legs.

Size Shoulder height 65–100 cm.
Weight 35–80 kg.

Habitat and Behaviour Animals of arid and rugged hill country with sparse vegetation cover, Nubian Ibex both browse and graze. They form small herds that usually number less than ten individuals. Dominant rams circulate their ranges during the mating season, checking on the breeding condition of ewes and moving on if no oestrus ewes are present. Rams may remain with a harem beyond the mating period, or wander away. Breeding is thought to be seasonal – in some areas kids are born between March and April. Normally a single kid weighing 2–4 kg may be dropped after about a 165-day gestation.

Distribution and Status The Nubian Ibex is rare throughout its range. There are very low numbers in the Red Sea Hills from northern Eritrea, Sudan and through Egypt and into Sinai. It occurs in Israel and Palestine, Jordan, the southern coastal mountains of Oman, Yemen, and central and northern Saudi Arabia. No population estimates are known, but there are probably fewer than 5,000 surviving in the region.

Conservation Areas Jabal Musa WR, Jabal Katrina WR (Egyptian Sinai); Avdat Canyon NP (Israel, and at least 15 other conservation areas); Wadi Mujib WR, Dana WR, Jebel Masadiwe (Jordan); Yalooni AOS, Jabal Samhan NP (Oman); At Tubayq R, Hawtat bani Tamim R (Saudi Arabia).

ARABIAN TAHR *Hemitragus jayakari*

Identification The Arabian Tahr is a rather small goat with a very limited distribution. Both sexes carry short, stout, back-curved horns (those of females are lighter). Unlike in other goat species, the male does not have a beard. Tahr have coarse, rather shaggy coats. The back and flanks are light sandy-brown in colour and the underparts are off-white. Males have a distinct dark dorsal crest and females have a paler crest. Facial markings are distinctive, with a broad black blaze across much of the face and a black stripe running from eye to mouth corner, and a paler stripe separating these two dark markings. During the winter months the coat is less sleek and shaggier. The tail is very short.

Size Shoulder height 60 cm.
Weight 30 kg.

Habitat and Behaviour Tahr occupy rugged hill country up to 1,800 m (5,900 ft) above sea level, with steep slopes, bare cliffs and cavities to shelter from the sun. They are grazers and browsers taking a range of plant foods. Unlike many arid area species, Tahr need regular access to drinking water, which makes them particularly vulnerable to hunters. They are often solitary or live in small family herds attended

Arabian Tahr ewe and young

Male has heavier horns than female and a distinctive dark cape

by one adult male. They are territorial, range over relatively small areas and the males mark their territories. Both sexes stamp their feet and emit sharp nasal whistles when alarmed. Tahr are amazingly agile in even the steepest terrain.

Distribution and Status The Tahr is restricted to the northern Hajar mountains of Oman, to (remnant populations) Jebel Haffit, Abu Dhabi, UAE, and to the Shimaliyah range in the north-east of the UAE. It receives little active protection in the UAE, but is protected in the Omani range. There are perhaps 2,000 surviving Tahr in Oman and possibly fewer than 50 in the UAE.

Conservation Areas Wadi Sarin-Jabal Aswad, Wadi 'Asya, Jabal Nakhl, Al Hamya, Jabal Bani Jabr (Oman).

SHEEP

BARBARY SHEEP *Ammotragus lervia*

Identification This is the only wild sheep species occurring in North Africa, and it is endemic to that region. A large, stocky sheep, the ram is much larger than the ewe. A well-developed fringe of long hairs – particularly noticeable in the ram – extends down the throat, chest and front legs. The species has a fairly long and well-haired tail, and short ears. The horns of the ram are particularly well developed, being stout at the bases and curved back, down and forwards. The animal is an overall sandy to reddish-brown colour, although the long fringe is usually paler. All six races described across the wide range of this species are readily identifiable as Barbary Sheep.

Size Shoulder height 75–120 cm.
Weight 40–140 kg.

Barbary Sheep ewe and lamb
Barbary Sheep have extensive throat fringes

Habitat and Behaviour Barbary Sheep occupy arid hill and mountain country across the Sahara and its fringes. They are mixed feeders and both browse and graze. They usually live in small flocks of up to 20 individuals. An adult ram accompanies the ewes and lambs; solitary rams are not uncommon. Mating peaks in October–November in some areas, with lambs being dropped after 150–65 days. The usually single lamb weighs 1.5–3 kg at birth; twins and triplets have been recorded.

Distribution and Status The Barbary Sheep once occurred through-out the Sahara where habitat was suitable, from the Atlas ranges to the Sudanese Red Sea Hills. Hunting and competition with livestock has caused local extinctions of this species in a number of areas, and has greatly reduced remaining populations. Population size is unknown, but estimates put it at less than 30,000. Substantial populations are said to survive in the massifs of Aïr, Termit and Ahaggar.

Conservation Areas Toubkal NP, Eastern High Atlas NP (Morocco); Belezma NP, Tassili n' Ajjer NP, Ahaggar NP, Djebel Aissa SF (Algeria); Djebel Chambi NP, Djebel Bou Hedma NP, Dghoumes NP (Tunisia); Zada-Archei FR (Chad); Aïr and Tenere NR (Niger).

WILD SHEEP *Ovis orientalis*

Identification Classification of the Wild Sheep is complex and confused, but it can be divided roughly into the Urial and Mouflon groups. In the region, especially Iran, five subspecies are recognized: the Transcaspian Urial (*O. o. arkal*) and Afghan Urial (*O. o. cycloceros*) from the east and north-east; and the Armenian Mouflon (*O. o. gmelini*), Esfahan Mouflon (*O. o. isphahanica*) and Larestan Mouflon (*O. o. laristanica*) from central and western areas. Interbreeding apparently takes place where ranges meet. The only race to occur outside Iran is the Armenian Mouflon (in extreme eastern Turkey), and an isolated western population that is sometimes considered a separate subspecies. It is possible that some animals may survive

Mouflon ram in partial moult – note the massive horns

Larestan Mouflon ram

Afghan Urial ram showing dark throat fringe

in the Kurdish region of northern Iraq. Animals hunted in the 20th century in Oman were probably introduced from Iran and it is doubtful if any survive.

All races are recognizable as Wild Sheep, with rams carrying massive, curved, forwards-facing horns that in some races may exceed 1 m around the curve. Ewes may be hornless or carry short, thin horns. The coats of Wild Sheep are short and usually sandy-yellow in summer, longer and somewhat darker in winter. Unlike rams of most goat species, Wild Sheep rams do not have beards. Urial rams have a bib of longish white to grizzled hair on the throat. Some Mouflon-race rams have a light-coloured saddle on the back and a dark-coloured bib on the lower neck (present in two races). Wild Sheep rams are larger than the ewes.

Size Shoulder height 50–100 cm.
Weight 25–85 kg.

Habitat and Behaviour Preferred habitats include hill and mountain country, but avoiding steep slopes and cliffs, and adjoining plains. Wild Sheep are principally grazers, taking grasses and herbaceous plants, but they do browse occasionally. They live in flocks of varying sizes. Outside the breeding season, rams run in bachelor herds. The rut occurs mainly in the autumn and early winter, with births taking place in the following spring. The number of lambs per ewe varies from one to three.

Distribution and Status See opposite.

Conservation Areas Uromiyeh Lake NP, Khogir NP, Sorkheh Hesar NP, Gamishlo WR, Kolahgazy WR, Khabr-va-Rochon WR, Bamou NP, Tandoreh NP (Iran); Konya-Bozdag R, Van-Özalp PBA (Turkey).

CAMELS

DROMEDARY *Camelus dromedaries*

Identification The Dromedary or One-humped Camel is present in this guide because it occurs across much of the region – it is frequently seen roaming and feral populations exist. No one is sure when it became extinct as a truly wild animal, but it is estimated that this happened between 2,000 and 3,000 years ago. The species probably originated in the Arabian Peninsula. Its size, long neck and legs, and the large hump on its back make it unmistakable. It varies greatly in colour, but most individuals are shades of brown or grey-brown.

Size Shoulder height 1.8–2.3 m.
Weight 300–690 kg.

Dromedary cow with twins

Habitat and Behaviour This is an animal of the desert and its fringes, a habitat to which it is particularly well adapted – it has large, rounded feet, nostrils that can close and the ability to go for long periods without water, drawing its needs from plant food. It grazes and browses, and can eat plants with a high salt content that other herbivores will not touch. Left to their own devices, camels may form into three kinds of herds: bachelor groups, females with newborn calves and females with young older than one year. A single adult bull that will drive off other approaching males leads all groups. The peak birth season is February–May over much of the range, but in north-east Sahara it is January–March. Calves have an average weight of 37 kg at birth.

Distribution and Status Found across the region.

EVEN-TOED NON-RUMINANTS (Suiformes)

Pigs are even-toed non-ruminants that have similar hoofs to antelopes, but they do not chew the cud. Unlike antelopes and deer, pigs have upper incisors.

EURASIAN WILD PIG OR WILD BOAR *Sus scrofa*

Identification This is a large typical pig that is covered with coarse, bristle-like, grizzled-brown to grey-brown hair. The piglets are dark brown with distinctive yellowish-brown longitudinal stripes. The snout is long, and although the tusks (canine teeth) are relatively short, in adults they protrude beyond the lips. The lower tusks of boars may reach 25 cm or more in length. Boars are considerably larger than sows.

Size Shoulder height 70–90 cm.
Weight 60–130 kg.

Habitat and Behaviour In their north-west African range, Wild Pigs occupy mainly oak and other forests; scrubland and small, isolated thickets are also utilized. They avoid semi-desert areas with little cover. In Turkey and Iran they occupy mainly broadleaf woodlands (especially areas dominated by oak trees), marshes with reed beds, dense scrub and riverine thickets. Wild Pigs are omnivores that eat a wide range of plant and animal foods, although the former usually dominate and include leaves, tubers, bulbs, fruits and seeds. In some areas they are regarded as a nuisance on cultivated lands. They are nocturnal, but if not disturbed or hunted they may be active in the cooler daylight hours. They live in groups or sounders of 6–20 individuals, which include one or more adult sows. Boars retain a loose association with these sounders. In Iran and probably elsewhere in the region, piglets are born in early spring. The sows may give birth to 4–10 young, and birth weight depends on litter size. Gestation is about 112 days.

Eurasian Wild Pigs are variable in coat colour

Piglets of Eurasian Wild Pig are striped when small

Distribution and Status Populations are present in and adjacent to the Atlas ranges of north-west Africa (Morocco, Algeria, Tunisia), Israel and Palestine, possibly Syria and Iraq, and wherever there is suitable habitat in Turkey and Iran. In Iran, major populations occur in the forests of the Zagros and Alborz ranges. Their best protection in the region is religious strictures laid down against eating pork.

Conservation Areas Present in many reserves throughout its range in the region.

Similar Species One other pig species, the Warthog (*Phacochoerus africanus*), is present in the region as an introduced species in far north-eastern Algeria. Isolated, naturally occurring populations are found in Niger, close to the southern border of Algeria. The Warthog has a typical pig-like appearance with grey, sparsely haired skin, a mane of long, erectile hair on the neck and back, wart-like lumps on the face and curved, upwards-pointing tusks. The thin tail is held erect when running. It is generally smaller than the Eurasian Wild Pig, and is active during the day.

ODD-TOED UNGULATES (Perissodactyla)

A single hoof (third toe) on each limb and well-developed incisor teeth on the upper and lower jaws characterize these mammals.

ONAGER *Equus hemionus*

Identification This is the only wild member of the horse family occuring in the region. It bears more resemblance to a donkey than a horse, hence its alternative name of Persian Wild Ass. The coat colour varies from grey-brown to yellowish, with a black stripe running down the centre of the back. The lower flanks and underparts are white, and the rump is usually pale to white. The mane is short and dark in colour. Feral donkeys occur in some areas of the Onager's distribution and could be confused with them from a distance.

Size Shoulder height 1.4 m.
Weight 150–250 kg.

Habitat and Behaviour The Onager is a species of arid and semi-arid plains, where it is mostly a grazer. Most feeding takes place during the hours of daylight, but this may vary according to levels of disturbance. It requires regular access to drinking water. Herd size is variable, but usually small because numbers have been drastically reduced by hunting and destruction of habitat by domestic stock. Mating takes place in June, when stallions fight viciously over territories and mares. The single foal is born after a gestation of about 11 months, and is suckled for at least a year.

Distribution and Status In historical times the Onager occurred from the eastern Mediterranean coast and eastern Turkey in an almost continuous belt through northern Arabia and across Iran and beyond. Today only a few hundred animals survive within the region, in Iran. The last Onager in Syria was noted in 1927, and the species disappeared elsewhere much earlier. Animals have been reintroduced in Israel.

Conservation Areas Khar Turan NP, Bahram-e Gur PA (Iran); Hai Bar NR, Maktesh Ramon NR (Israel).

Similar Species The African Wild Ass (*Equus asinus*), the ancestor of the widely distributed domestic donkey, once inhabited parts of Arabia (although probably not after the Neolithic period). However, other than in its domestic form it no longer occurs in the region. It survives only in very small numbers in a limited area in the Horn of Africa.

HYRAXES (Hyracoidea)

These small social mammals are related to ungulates; their two upper incisors grow throughout life (as these teeth do in elephants, for example).

ROCK HYRAX *Procavia capensis*

Identification A large amount of controversy surrounds the taxonomy of the Rock Hyrax in the region – here this one species is referred to. Rock Hyraxes vary greatly in coat coloration even within the same population, and this is often the criterion used to claim species and subspecies status. Hyraxes are small and stout with no visible tail and short, rounded ears. The coat colour ranges from greyish-fawn through yellowish-fawn to dark brown; the underparts are slightly paler. A patch of erectile hair is located in the centre of the back surrounding a scent gland. The colour of this hair varies in different areas and populations, but in the region it is usually yellowish-buff or has a reddish hue; in some populations it is dark brown to black.

Size Total length 54 cm.
Weight 2.5–4.6 kg.

Habitat and Behaviour This is a species of rocky habitats including mountains, rocky outcrops, gorge walls and sea cliffs. Rock Hyraxes are predominantly diurnal, but they also feed during warm moonlit nights. They spend a great deal of time basking in the sun at the

Hyraxes are sociable animals that live in small colonies

entrances to their crevice homes. Hyraxes eat a wide variety of plants, including species that are poisonous to other species. They live in small colonies of 4–8 animals. In optimum habitat this may rise to 20 or more individuals, and many groups may live in close proximity to each other. Each group has a dominant male and female, and there is a clearly demarcated social ranking. The young are well developed at birth and soon able to move around. One to three cubs are dropped after an unusually long gestation of about 210 days. Birth weight is 150–300 g.

Distribution and Status Hyrax populations occupy all the major massifs in the Sahara, including Ahaggar, Tibesti and Tasilli n'Ajjer. Hyraxes are absent from the Atlas ranges. In the east they are found through the Sudanese and Egyptian Red Sea Hills into Sinai. They occur in Israel and Palestine, Lebanon, Jordan and right through the western hills and mountains of Saudi Arabia and Yemen. Isolated populations live in the coastal hills of Dhofar in southern Oman. Hyraxes do not occur naturally in the UAE, but a population has been introduced to Jebel Haffit, Al Ain.

Conservation Areas Ahaggar NP, Tasilli n'Ajjer NP (Algeria); Aïr-Ténéré NR (Niger); several in Israel; Jebel Samhan NP (Oman).

CARNIVORES (Carnivora)

Some 35 species of carnivore inhabit the region, including nine members of the dog family (Canidae) and nine of the cat family (Felidae). The Lion and Tiger once occurred widely, but both species were driven to extinction in the mid-20th century (see page 8). Remaining large carnivore species are under threat due to habitat loss and direct persecution by man. The teeth of carnivores are specialized for killing and eating other animals.

HYAENAS

STRIPED HYAENA Hyaena hyaena

Identification As is typical of all hyaenas, the shoulders of the Striped Hyaena stand higher than the rump. The shaggy, buff to grey coat is covered by numerous transverse black stripes. A well-developed erectile mane extends from the base of the neck to the rump, and there are numerous black stripes and rings on the legs. The head is large, much of the muzzle is naked and the throat area is mostly black. The ears are long and pointed, and the tail is relatively long and bushy.

Size Shoulder height 72 cm.
Weight 40–55 kg.

Habitat and Behaviour This hyaena shows a strong preference for dry areas, often in association with rocky outcrops and within savannah. Records are few from the central areas of the major deserts, although

The Striped Hyaena has a long, erectile cape

hyaenas are present in and around the major massifs such as Ahaggar Tassili and Aïr. In some areas this species is also found on coastal plains, especially in southern Morocco (Western Sahara) and Oman. It is an opportunistic feeder, taking a wide range of animal and plant foods. In some areas it actively hunts larger prey such as sheep and goats, and there are a number of documented cases of children being killed and eaten by it in India. It is also a scavenger, and in some areas is a regular visitor to town rubbish dumps. It is reported to dig up human corpses from freshly dug graves.

The majority of Striped Hyaena sightings are of solitary animals and occasionally pairs. Most activity takes place at night. Records indicate that in some areas hyaenas hunt in groups. They probably live in loosely associated groupings within a shared home range, but usually forage alone. In some areas births show seasonal peaks, but in general births seem to be aseasonal. After about a 90-day gestation, 2–4 cubs are dropped in a rocky den or a burrow dug by another species.

Distribution and Status The Striped Hyaena occurs at low densities across most of the countries covered by this guide. It is apparently absent from much of the north-west and western provinces of Iran but occurs in south-east Anatolia and the Turkish–Syrian border region. These seem to be the only parts of Turkey where it is still found. In the Arabian Peninsula it has been heavily persecuted but still survives in parts, associated especially with the coastal plain and adjoining mountain ranges.

Conservation Areas Ahaggar NP, Tassili n'Ajjer NP (Algeria); Aïr and Tenere NR (Niger); Banc d'Arguin NP (Mauritania); Dasht-e Arjan PA, Lake Bakhtegan WR, Hamoun-I Sabari PA, Lower Sarbaz River PA, Naybandan WR (Iran); Azraq Oasis WR (Jordan); Jabal Samhan N (Oman); Raydah Escarpment NR (Saudi Arabia).

The Aardwolf is much smaller than the Striped Hyaena

Similar Species In the Red Sea Hill region of Sudan and south-east Egypt, the Aardwolf (*Proteles cristatus*) is said to occur on the basis of a number of old records. It could be confused with the Striped Hyaena. The Aardwolf is the smallest hyaena. It is similar in appearance to the considerably larger Striped Hyaena, with a shoulder height of 50 cm; weight of up to 11 kg; pointed, fairly large ears; body and leg striping; and mane of erectile hair down the neck and back. It also has well-developed canine teeth, but the remainder of the teeth are few and rudimentary. It almost exclusively eats termites and to a lesser extent ants.

CATS

CHEETAH *Acinonyx jubatus*

Identification The 'greyhound' of the big cats, the Cheetah has a slender body and long legs, with a long, black-spotted and ringed, white-tipped tail. Its head is small and short-muzzled for its size, with a black line running from the inner corner of each eye to the corner of the mouth. This 'tear-line' is unique to the Cheetah. The overall body colour is off-white to pale fawn, liberally sprinkled with more or less uniformly sized, rounded black spots. Animals living in the Sahara tend to be paler than those elsewhere, and have shorter coats. Where snow falls in winter in parts of their Iranian range, Cheetahs' coats tend to grow longer at this time. Young cubs have an extensive mantle of longish grey hair.

Size Shoulder height 80 cm.
Weight 30–72 kg.

Cheetah, showing dark line between eye and corner of mouth

Habitat and Behaviour The Cheetah's preferred habitat is open savannah and semi-desert country, but in desert areas it is often found in association with hill and mountainous terrain. Unlike other cats, which stalk and pounce on prey, the Cheetah relies on sight and speed (hence the open nature of the habitats that it prefers). It sprints in for the kill at up to 70 km per hour, but this speed can only be maintained for a few hundred metres. Cheetahs favour medium-sized mammal prey with a weight of up to 60 kg. The North African population includes gazelles, sheep, goats and young camels in its diet. In Iran Cheetahs hunt gazelles, wild sheep, wild goats and hares, but because of the reduced prey base they also kill domestic sheep and goats. Most hunting takes place during the day.

Cheetahs are normally solitary, but females may be encountered with dependent cubs, and males may form a coalition. Cooperative males are usually from the same litter. Females establish territories from which they will drive out other females. Heavy human hunting

A rare photograph of a Cheetah taken in Iran

pressure has drastically reduced Cheetah populations, and this probably has a major influence on social structure. Litters of 1–5 cubs weighing up to 300 g at birth may be born at any time of year. The cubs remain hidden for at least their first six weeks of life.

Distribution and Status Cheetahs once occurred across much of North Africa and Arabia, and through Iran, into Pakistan and India. Today they occupy only a fraction of their former range in south and eastern Algeria and adjacent areas of Niger and Libya, centred on the Ahaggar massif. It is likely that remnant populations survive elsewhere on the southern Saharan fringe. The Asian population (*A. j. venaticus*) survives in isolated areas of north-east and east-central Iran. Estimates vary, but only 60–80 cheetah occur in that country.

Conservation Areas Ahaggar NP; Tassili n' Ajjer NP (Algeria); Aïr and Tenere NR (Niger); Naybandan WR, Ariz and Bafq PA, Daranjir WR, Kavir NP, Khar Turan NP (Iran).

LEOPARD *Panthera pardus*

Identification This elegant and powerfully built species of cat occurs widely in the region in very low numbers. Its body colour varies from off-white to orange-russet with black spots on the lower legs, flanks, hindquarters and head. Spots over the rest of the body consist of rosettes or broken circles of irregular spots. The white-tipped tail is about half of the total length of the animal. The underparts are usually paler than the upperparts. Unlike in the Cheetah, there is no black 'tear mark' between the eye and mouth corner. Leopards vary greatly in size from area to area: smaller animals occur in the Arabian Peninsula; among the largest in the mountains of Iran.

51

Size Shoulder height 70–80 cm.
Weight 17–90 kg.

Habitat and Behaviour Leopards live in a wide range of habitats, avoiding only open desert country, but mostly remain in rugged hill and mountain areas in the region. They have survived where other big cats no longer occur due largely to their catholic diet. This ranges from insects, reptiles and fish to birds, but the bulk of their prey is made up of small to medium-sized hoofed mammals, especially antelopes. In some areas Wild Pigs are important in their diet. Hyraxes and hares are also preyed upon in parts of the region, as are goats, sheep, young camels and horses where natural prey has been depleted; they will also readily scavenge.

The Leopard is a solitary cat, with the exception of a mating pair that comes together only for a short period, or females accompanied by cubs. Because of disturbance and hunting pressure, Leopards tend to be nocturnal, secretive and seldom seen. Adult males mark and defend a territory against other males, and a male range may overlap the ranges of several females. Territories are marked with urine scrapes, droppings, tree scratching and the distinctive deep 'sawing' or grunting call. Home range size is influenced by prey abundance and availability. The cubs weigh about 500 g and usually number 2–3 in a litter. They may be born at any time of year after a gestation of about 100 days.

Distribution and Status Leopards have been greatly reduced in number across the region, but still occur in parts of the Atlas ranges, the Ahaggar and Tassili N'Ajjer massifs in southern Algeria, the western Red Sea Hills, Israel and Jordan, and the coastal mountain ranges of the Arabian Peninsula, as well as in south-western (Taurus Mountains) and far eastern Turkey. They have a wide but patchy range in Iran.

Conservation Areas Ahaggar NP (Algeria); Aïr and Tenere NR (Niger); Judean Desert NR, North Arava Valley NR (Israel); Jabal Samhan NP (Oman); Al Fiqrah PA (Saudi Arabia); Ariz and Bafq PA, Kiamaki WR, Lisar PA, Parvor PA, Khosh-Yeilagh WR, Tauran WR, Sarani PA, Tandoureh NP, Oshtrakuh PA, Baman NP (Iran).

WILD CAT *Felis silvestris*

Identification The Wild Cat has a wide distribution across the region, with three distinct groups of races recognized: the European (*F. s. silvestris*), Asiatic (*F. s. ornata*) and African (*F. s. lybica*). The appearance and form of the species is similar to that of the Domestic Cat, but it is generally longer in the leg and larger. It hybridises readily with domestic and feral cats, which can make **identification** confusing. Reddish-brown to yellowish fur is often present on the belly and inner surfaces of the hind legs. The overall body colour ranges from pale sandy-brown to light or dark grey. Paler animals are usually found in low rainfall areas. The coat length also varies according to prevailing climatic conditions in different areas. The body is marked to a greater

or lesser extent with darker vertical stripes, and the eastern form also has dark spots. Sometimes dark rings are present on the legs. The longish dark-ringed tail has a black tip. The chin and throat are white, and the underparts are paler than the upperparts.

Size Shoulder height 35 cm.
Weight 2.5–6 kg.

Habitat and Behaviour Wild Cats occupy virtually all habitats, but are generally absent from the heart of true desert (except where there is hill country). They hunt a wide range of prey. Small rodents are particularly important, but they also take young hares, hyraxes, young gazelles, birds, and more rarely reptiles and insects. They are solitary, except during the brief mating period and when young accompany a female. Mainly nocturnal and crepuscular, they lie up during the day in rock crevices, trees, dense vegetation or burrows dug by other species. Both sexes establish, mark and defend territories. Wild Cat densities can be quite high in areas of optimal habitat and abundant prey. After a gestation of 56–65 days, 1–5 kittens are born, each weighing just 40–50 g. Breeding takes place in the Sahara in January–March. There are variations of breeding information for the different groups.

Distribution and Status Although Wild Cats are hunted across much of their range they are still abundant. The main threat is interbreeding with domestic and feral cats. The African group occurs around the fringes of the Sahara and across to the coastal plains, with populations located on all the major massifs within the desert. This group is found all around the Arabian Peninsula coastline and adjacent mountain ranges. The European group occurs across northern and much of southern Turkey. The Asian group lives across Iran to the east of the Zagros Mountains.

Conservation Areas The Wild Cat is found in many reserves across its very wide range.

JUNGLE CAT *Felis chaus*

Identification Also called the Swamp Cat, this is a very widely distributed cat that lives from the Egyptian Nile eastwards to Vietnam. Although similar in overall appearance to the Wild Cat, it is larger and not as distinctly marked. As in a number of Wild Cat races, the backs of the ears are reddish-brown with a short tuft of black hairs (to 15 mm) on the ear tips. The tail is proportionally shorter than that of the Wild Cat. The overall body colour is pale sandy-brown, with indistinct markings on the sides and legs. The underparts are paler.

Size Shoulder height 45 cm.
Weight 2.5–12 kg.

Habitat and Behaviour The Jungle Cat is usually found in wetland associations with good cover such as reed beds, riverine thickets, forest and agricultural land adjacent to these habitats. It is generally

solitary, but male, female and kitten groups have been reported. Jungle Cats are nocturnal and diurnal hunters of rodents, and also take hares, birds, reptiles and fish. Young deer and Wild Pigs are also killed. Unusually for cats, Jungle Cats are reported to swim well and will dive to catch fish. Three to five kittens are born after about a 66-day gestation, with birth weights of up to 55 g each.

Distribution and Status In North Africa the Jungle Cat is only found in the Nile Delta and on the floodplains of the river within Egypt. Absent from the Arabian Peninsula, it occurs in Israel and Palestine, Lebanon, northern Syria, patchily in Turkey, along the Euphrates and Tigris Rivers in Iraq and in many areas where suitable habitat is present in Iran, such as the Caspian forests.

Conservation Areas Golardi Sulun, Karatepe-Aslantas (Turkey); Anzali Mordab WR, Amirkelayeh Lake WR, Miankaleh Peninsula WR, Karkheh River Marshes WR, Dez River Marshes WR, Dasht-e Arjan PA, Lake Bakhtegan WR (Iran).

SAND CAT *Felis margarita*

Identification This is the smallest cat in the region, with a somewhat broadened face and large, rounded ears. The fur is thick, soft and more or less uniformly pale sandy-buff, with indistinct darker banding on the body and legs. The face is usually paler than the rest of the body, and there are dark spots on the backs of the ears. The underparts are off-white in colour. Long hair grows between the toe pads, an adaptation to the substrate on which the cat lives.

Size Shoulder height 30 cm.
Weight 1.5–3.5 kg.

Habitat and Behaviour The Sand Cat is a predator of sandy desert, including extensive dune country and also stony areas. It is mainly nocturnal and solitary, living in burrows excavated under shrubs and tussocks. It is one of the few cats known to dig for prey and shelter. Rodents make up much of its diet, but it also eats reptiles, birds and invertebrates. Births have been recorded in January–April in the Sahara. Two to five kittens, each weighing 55 g, are dropped after a 63-day gestation.

Distribution and Status The Sand Cat probably occurs across much of the Sahara to the Red Sea Hills. It is widespread in the Arabian Peninsula, with a limited range in Israel and Palestine, Jordan and Iraq west of the Euphrates River. In Iran it appears to be restricted to the north-east and Baluchestan in the east.

Conservation Areas Tassili N'Ajjer NP, Ahaggar NP (Algeria); Bou-Hedma NP (Tunisia); Aïr and Tenere NR (Niger); Hai Bar NR, North Arava Valley NR (Israel); Mahazat as-Sayd NR, Harrat al-Harrah NR (Saudi Arabia); Jal Az-Zor NP (Kuwait); Kavir NP (Iran).

PALLAS'S CAT *Felis (Otocolobus) manul*

Identification Also called the Manul or Steppe Cat, this stocky, long-haired cat is limited to Iran within the region. The hair on its underparts and tail is twice as long as the hair on its back and sides, an adaptation to the cold nature of its chosen habitats. Its fur colour varies, but within Iran it is usually buff with fox-red infusion. The hairs are tipped with white, lending a somewhat silvery appearance. The short legs are marked with indistinct black bands, and the tail is black-tipped with a few black rings. The cat has a large, flattened head and short, rounded and low-set ears. The black spotting on its forehead and black stripes on its cheeks are diagnostic. The Sand Cat and Pallas's Cat both have a characteristic barking call.

Size Shoulder height 25–30 cm.
Weight 2–4.5 kg.

Habitat and Behaviour This is a solitary cat of stony alpine desert and open high-altitude grassland and shrub country. It is especially prevalent where there are rocky outcrops and scree. It has not been studied in Iran, but elsewhere it hunts rodents, pikas and also occasionally birds. It is mainly active around dusk and dawn, but because some of its most important prey species are active during the day, this indicates that it must also hunt at this time. When not active Pallas's Cats den down in caves, rock crevices and burrows dug by other species. Most litters are apparently born in spring (April–May), and comprise 1–8 kittens, each weighing 80 g.

Distribution and Status The species is restricted to northern Iran, particularly to the north-east, the Alborz ranges and the north-western provinces of Azarbaijan, where it said to be very rare. It has a very wide north Asian distribution. Unfortunately, it is heavily hunted for its pelt and is considered seriously threatened in Iran.

Conservation Areas None known.

EURASIAN LYNX *Lynx lynx*

Identification The Eurasian Lynx has relatively long legs and large paws. The coat colour is predominantly greyish with a tinting of yellowish to bright red and profuse black spotting. The short tail (about 18 cm) is black-tipped. Prominent tufts of longish hair hang down from the cheeks. The ears are long and pointed with black tufts (35 mm long) at the tips, and there is a white patch at the centre of the back of each ear.

Size Shoulder height 60–75 cm.
Weight 12–22 kg.

Habitat and Behaviour In the region the Eurasian Lynx is restricted to forested mountains and hill ranges in Iran and Turkey. Roe Deer and Chamois are said to be its main prey, but it also takes young Red Deer, Wild Sheep and Wild Goats, as well as hares and large rodents. It is not unusual for a Eurasian Lynx to kill prey up to three times its own weight. It is a solitary hunter that is active mostly in the morning hours and evening. Over much of its range young are born in May–June after an average 69-day gestation. Litter size is 1–4 kittens, which are independent of the mother at about 10 months.

Distribution and Status The Eurasian Lynx is very rare in both Turkey and Iran mainly because of habitat and prey destruction, but also due to direct hunting by man. In Turkey it is mainly restricted to the forested mountain areas. In Iran it is only known from the Alborz range south of the Caspian, and it has recently been recorded in the Zagros Mountains in the west.

Conservation Areas Munzur NP (Turkey); Dena PA, Kiamaki WR, Arasbaran PA, Lisar PA, Kavir NP, Golestan NP (Iran).

CARACAL *Caracal caracal*

Identification The Caracal is a powerful medium-sized cat with hindquarters that are slightly higher than its shoulders and a short, uniformly coloured tail. The coat ranges from uniform pale yellow-fawn to a rich brick-red. The underparts vary from slightly paler than the upperparts to off-white, and in some individuals faint spotting may be present. The face is marked with black and white patches. The Caracal's long, pointed, dark-backed (with white sprinkling), black-tufted ear tips are diagnostic.

Size Shoulder height 40–45 cm.
Weight 7–19 kg.

Habitat and Behaviour This species occurs widely, although patchily, throughout the region in a variety of habitats, but avoids true desert. Caracals are mainly nocturnal, but in areas where they are not hunted or disturbed they also hunt in the cooler daylight hours. In areas where diurnal hyraxes form a large part of their prey, Caracals are forced to hunt during the day. They prey on mainly small to medium-sized mammals weighing up to approximately 40 kg, but also take birds and more rarely reptiles. They are heavily persecuted in areas where they prey on domestic sheep and goats. Normal litter size consists of 1–3 kittens, each weighing about 250 g, and dropped after an average 79-day gestation.

Distribution and Status Caracals occur across the whole of the northern Sahara, the Sahel to the south and extending to the Atlantic Ocean and Mediterranean and Red Sea coastlines, with populations associated with the main Saharan massifs. They are found in the mountain and hill ranges of Arabia, as well as in some parts on to the coastal plain. They are present in Israel and Palestine, Jordan and Iraq,

but only occur in the far south of Turkey. In Iran this cat is associated mainly with eastern and central areas, but it has been recorded throughout the country.

Conservation Areas Tassili N'Ajjer NP, Ahaggar NP (Algeria); Aïr and Tenere NR (Niger); Fada Archei FR (Chad); Judean Desert NR, North Arava Valley NR (Israel); Azraq Oasis WR (Jordan); Raydah Escarpment NR (Saudi Arabia); Dilek Yarimadisi (Turkey); Khosh-Yeilagh WR, Tauran WR, Sarani PA, Dasht-e Arjan PA (Iran).

SERVAL *Leptailurus serval*

Identification The Serval is similar to and often mistaken for the Cheetah and even Leopard, but it is much smaller than either species and has a short, black-ringed tail and large ears. The body is usually yellowish-fawn and there are numerous black spots across the body and on the legs.

Size Shoulder height 60 cm.
Weight 8–13 kg.

Habitat and Behaviour In its very limited North African range, the Serval frequents humid scrub and mixed woodlands. It is active during both the day and night, hunting singly, in pairs or in small family parties. Rodents make up the bulk of its diet. There are usually 1–3 kittens to a litter, which are dropped after a gestation of about 73 days, with each kitten weighing 200 g. Almost nothing is known about this cat within the region.

Distribution and Status The Serval is widespread and sometimes common in sub-Saharan Africa, but in the north it has a very limited distribution. Relict populations in the northern Atlas ranges of Morocco, Algeria and Tunisia are endangered and may be extinct.

Conservation Areas None known.

D FOX *Vulpes vulpes*

entification This is the largest true fox in North Africa and the ddle East. It has a long, white-tipped, bushy tail. The ears are ge, dark on the outer surfaces and white on the inner sides. The at colour is greyish-russet to yellowish-grey, with a darker dorsal nd running from the tail base to the head. The underparts are ually paler than the upperparts. The lips, chin and throat are pale to ite. Seven subspecies occur within the region; all should be readily cognizable as the Red Fox.

ze Shoulder height 30 cm.
eight 4–10 kg.

abitat and Behaviour The Red Fox avoids true sand desert and high ountain ranges, but otherwise occupies a wide variety of habitats. is usually common wherever it occurs, and can often be found se to human habitation and agricultural land. Its wide range and undance is to a large measure due to its extremely varied diet. It kes both animal and plant foods, readily scavenges, and raids rubbish mps and dustbins. In areas where there are date plantations, it ts the fallen fruits of the palms. Red Foxes dig burrows or modify ose excavated by other species. A mated pair usually establishes d defends a territory, which it may share with its non-breeding fspring. Most activity takes place at night, but Red Foxes also forage ring the cooler daylight hours. Three to twelve pups, each weighing –150 g at birth, are dropped after a gestation of 51 days.

stribution and Status The Red Fox is present throughout the Atlas nges and their fringes, along to the Egyptian coast and Nile Delta, d down the Nile River into central Sudan. It occurs some distance from the Nile, especially where there are hills. Otherwise it is found throughout the Middle East, although it is absent from the heart of the Great Sand Desert of the Arabian Peninsula.

Conservation Areas Virtually all throughout its range.

CORSAC FOX *Vulpes corsac*

Identification This species is similar to the Red Fox (*V. vulpes*), but considerably smaller with a black tail tip. The tail is about half the length of the head and body. The ears are fairly small compared with those of other regional foxes; they are either ochre-grey or reddish-brown on the back surfaces. In winter the Corsac Fox's fur is long and greyish straw-white in colour, but in summer it is short and buff-rufous. Its underparts are white to pale yellowish-white.

Size Shoulder height 30 cm.
Weight 2–2.5 kg.

Habitat and Behaviour Corsac Foxes occupy open steppe country and avoid areas with closed vegetation and mountain ranges. They are opportunistic feeders that will take a wide range of animal food, with small and medium-sized rodents making up the bulk of their prey, and will readily scavenge from carcasses and human refuse and waste. They are both nocturnal and crepuscular, and live as life-time mated pairs. Mating takes place during January–March, with the first births occurring in March and the majority in April after an estimated gestation of about 56 days. The newborn pups, 2–10 a litter, each weigh 60 g.

Distribution and Status In the region the Corsac Fox only occurs in the Gorgan-Turkoman steppe region in Iran, along the Turkmenistan and Afghanistan borders. Elsewhere it occurs throughout the high country of southern-central Asia. It is heavily hunted, especially for its winter pelt.

Conservation Areas None known.

BLANFORD'S FOX *Vulpes cana*

Identification This is a very small, attractive fox with a long and extremely bushy tail that is nearly always black-tipped (in about 20 per cent of the south-eastern Arabian population the tip is white). The body hair is soft, dense and brownish-grey in colour, fading to pale yellow on the belly and richer yellow-fawn on the face. The dorsal coat has a liberal sprinkling of white hairs, giving it a silvery appearance from a distance. Iranian populations are greyer than Arabian ones. The muzzle is sharply pointed with distinctive black markings, and the inner edges of the large ears are lined with white hairs. This is the only fox in the region with semi-retractile claws to aid climbing.

Size Shoulder height <25 cm.
Weight 1–1.3 kg.

Corsac fox

Blanford's fox

Habitat and Behaviour This is a fox of mountainous and rugged hill country (although it avoids high altitudes), and a strong preference for rocky and steep mountain slopes adjacent to gravel and sand plains (but onto which it seldom ventures). It is able to ascend and descend vertical and loose rock cliffs with amazing agility. Blanford's Foxes eat large quantities of wild and cultivated fruits, as well as considerable numbers of invertebrates. Small rodents and lizards are also hunted. This fox is strictly nocturnal and lives in monogamous pairs within relatively small, defended territories. Where it is not hunted or disturbed it readily enters camps. Litters of 1–3 pups, each weighing an average 29 g, are dropped after a 50–60-day gestation. Mating in Israel and Palestine takes place in January–February and the pups are born in early spring.

Distribution and Status Blanford's Fox has a wide distribution in central and eastern Iran, occurring up to the eastern Zagros ranges. Populations of the fox live in the coastal mountains of the UAE, Oman and Saudi Arabia, and probably in Yemen (although this needs confirmation). The fox is found in south-eastern Israel and Palestine, probably in Sinai and, although there is only one record for the Egyptian Red Sea Hills, its range almost certainly extends into north-eastern Sudan. It is a fox that is easily overlooked, and its range may still be found to be more extensive than is currently known.

Conservation Areas Jebel Samhan NP (Oman); Dana NR (Jordan); Ein Gedi NR, Judaean Desert NR, Maktesh Ramon NR, Eilat Mountain NR (Israel).

RÜPPELL'S FOX *Vulpes rüppelli*

Identification This is a small, lightly built fox with large ears. Its overall coat colour is pale fawn to almost white, usually with a russet tinge along the back, as well as on the face and the back of the ears. It has a long, bushy tail with a distinctive white tip.

Size Shoulder height 30 cm.
Weight 1.2–3.6 kg.

Habitat and Behaviour Rüppell's Fox shows a strong preference for sandy desert, including areas of dunes, but in some areas occupies stony desert. Its favoured areas have sparse to very sparse vegetation cover. It eats mainly invertebrates and small rodents, but also wild fruits and berries. It lives in territorial pairs that mate for life, but

occasionally larger numbers are seen together. These are probably family parties, with groups numbering 3–15 individuals. In Oman they occupy home ranges of about 50 sq km; elsewhere ranges may be as small as 10 sq km. The fox is mostly nocturnal with some diurnal activity. In Saudi Arabia mating takes place in December–February, with young cubs being found there and in Western Sahara in March. Litter size is reported from Saudi Arabia as usually 2–3 cubs.

Distribution and Status This small fox ranges right across the Sahara, but is absent from the Atlas region. It is also present through suitable habitat across the Middle East and the Arabian Peninsula, and across Iran, but is absent from much of Syria and does not occur in Turkey. In Iran it reaches its main distribution in the central deserts, the south and towards the Pakistan border.

Conservation Areas Ahaggar NP, Tassili n'Ajjer NP (Algeria); Sidi Toui NP (Tunisia); Nefhusa NP, Zellaf NR (Libya); Banc d'Arguin NP (Mauritania); Aïr and Tenere NR (Niger); Gebel Elba CA (Egypt); Maktesh Ramon NP, Tznifim NR (Israel); Mahazat as-Sayd, Harrat al' Harrah, Uruq Bani Ma'arid (Saudi Arabia); Jiddat al Harasis (Oman); Mehrouyeh WR (Iran).

PALE FOX *Vulpes pallida*

Identification This is a small, very pale fox with fairly long legs, large ears, a pale face and dark eye-rings. The body fur is creamy-white to sandy-fawn and sometimes exhibits a darker dorsal line. The underparts are generally paler than the upperparts, and there is pale rufous on the legs. The species sports a long, bushy, pale reddish-brown tail with a distinctive black tip.

Size Shoulder height 25 cm.
Weight 2–3.6 kg.

Habitat and Behaviour The Pale Fox occupies dry, sandy and stony areas in the semi-desert country of the Sahel and southern fringes of the Sahara. Almost nothing is known about it. The structure of its teeth indicates that it eats mostly plant foods such as fruits and berries, but it also catches small rodents, birds, reptiles and invertebrates. Pale Foxes normally associate in pairs and family parties, and are active at night, resting in burrow systems during the day. The extensive burrows may be as much as 15 m in length, and the inner chambers may be lined with dry plant material. Three to six pups are dropped after a 52-day gestation.

Distribution and Status In the region the Pale Fox occurs near the southern areas of Morocco (Western Sahara), Algeria and Libya, and it may well be found in these three countries given the availability of suitable habitat. It occurs in Mauritania, Mali, Niger, Chad and Sudan.

Conservation Areas None known.

FENNEC FOX *Vulpes zerda*

Identification This species is arguably the world's smallest fox, although Blanford's Fox falls within a similar weight range. The Fennec Fox is easily recognized by its extremely large ears (15 cm) and overall cream-coloured fur. The coats of many animals have a light fawn, reddish or greyish tinge, but the underparts are always paler. The foxes tend to be darker in the north of their range and much paler in the south. Dark streaks run from the inner eye down to either side of the muzzle. The tail is bushy and slightly darker at the tip. The footpads are covered with dense fur.

Size Shoulder height 19–21 cm.
Weight 1–1.5 kg.

Habitat and Behaviour This is a fox of sandy desert, including stable sand dunes as well as dunes near the Atlantic coast in Morocco (Western Sahara). It eats a wide range of both animal and plant foods, including insects, small rodents, lizards, small birds, wild fruits and some tubers. It is said to also forage around human settlements. Fennec Foxes are solitary hunters that obtain much of their animal prey by digging, with most activity occurring at night. Their large ears allow them to hear movement below the surface. During the day they shelter in burrows that can be complex, with up to 15 entrances in relatively compact sand. They live as mated pairs with pups of the season and even of the previous litter. Mating takes place in January–February, with births occurring in March–April. Litters may contain 1–5 pups, each averaging 45 g in weight.

Distribution and Status The nature of the Fennec Fox's chosen habitat is its best protection. It occurs throughout the Sahara, from the Atlantic almost to the west bank of the Nile, in the north in places to the Mediterranean and southwards to the northern Sahel Belt. It is found wherever there is suitable habitat in Mauritania, Mali, Niger, Chad and Sudan to the west of the Nile, and extends into the northern Egyptian Sinai Peninsula. Although two records exist from the north-eastern Arabian Peninsula (Kuwait and near Basrah, Iraq), its natural presence here needs to be confirmed. A number of early records were confused with the superficially similar Rüppell's Fox.

Conservation Areas Ahaggar NP, Tassili n'Ajjer NP (Algeria); Sidi Toui NP (Tunisia); Nefhusa NP, Zellaf NR (Libya); Bir El Abd CA (Egypt); Banc d'Arguin NP, Diawling NP (Mauritania); Aïr and Tenere NR (Niger).

GOLDEN JACKAL *Canis aureus*

Identification Also known as the Common or Asiatic Jackal, this is the only jackal to occur across North Africa and the Middle East. It has a typical jackal appearance, with variable coat coloration. The tail tip is always dark or black. The coat colour is commonly pale golden-brown with a liberal sprinkling of black and grey hairs on the back and sides – the colour may vary between individuals, populations and even seasonally. The head, backs of the ears and legs are often fawn to reddish, and the underparts are paler. Some animals have a fairly dark back saddle that could cause them to be confused with the Black-backed Jackal (*C. mesomelas*), but the closest region in which these two species overlap is in Eritrea.

Size Shoulder height 38–50 cm.
Weight 6–15 kg.

Habitat and Behaviour The Golden Jackal occupies a wide range of habitats from desert to forests, from sea level to 3,800 m (12,500 ft), and it is commonly seen in the vicinity of human settlements. It eats animals and plants, including young antelopes, hares, rodents, insects and carrion, as well as wild and cultivated fruits. Where food is abundant, up to 20 jackals may be found together, but these

gatherings are of a temporary nature. Foraging may be a solitary or group activity, with several animals cooperating in the hunting of larger prey. A pair mates for life and defends a territory, the size of which varies according to habitat and food availability. The pair tolerates the presence of the subadults from the previous litter as they help to feed and protect the current litter. Reproductive activity in Israel takes place in October–March. Timing is not known elsewhere in the region, but is probably similar, given that pups are born in February–March in Turkmenistan adjacent to north-eastern Iran. Usually there are 3–6 pups (ranges from 1–8) in a litter, dropped after a 63-day gestation. Jackals may be both nocturnal and diurnal: this is often determined by season and level of human disturbance.

Distribution and Status Golden Jackals occur throughout the region and in every country, only being absent from far northern Turkey and north-west Iran. In Iran they are said to be more common in the north than in the south. There have been local extinctions and this species may no longer occur in the UAE and parts of Oman. It is probably absent from much of the true desert.

Conservation Areas Present in many protected areas throughout its range.

GREY WOLF *Canis lupus*

Identification A large, unmistakable predator, the Grey Wolf occurs from the Egyptian Sinai eastwards within the region. Two subspecies are recognized in the region: the smaller *C. l. pallipes* (*C. l. arabs* is sometimes recognized for the Arabian Peninsula) occurs in the south. The larger, longer-haired *C. l. lupus* is found in the higher country of northern Turkey and Syria, and perhaps Iran. All, however, are obviously wolves, although young animals may be mistaken for Golden Jackals. At first glance wolves have the appearance of a German Shepherd Dog, except that they are longer in the leg and have large paws. Their fur is thick, except in animals from lower-lying hot areas, and usually predominantly grey or grey-brown. A yellow-sandy wash may be present.

Size Shoulder height 60–70 cm.
Weight 15–55 kg.

Habitat and Behaviour The Grey Wolf occupies a great range of habitats, and in the region is only naturally absent from the heart of the true deserts. Wolves are pack animals, with the size of groups largely dependent on food availability. The units usually consist of a family group, but solitary animals and pairs are encountered, for example in parts of the Arabian Peninsula. Although it has been claimed that wolves in Arabia seldom howl, this is not always the case. We have heard wolves howling in southern Oman and elsewhere. Favoured prey includes deer, antelopes, Wild Sheep and Goats, Wild Pigs and domestic stock where natural prey numbers have been depleted. Smaller prey such as hares and rodents, and even insects, is hunted, and wolves will also raid rubbish dumps. In packs only the dominant male and female usually breed. Pups are dropped in the spring after a gestation of 62 days. Although 1–11 pups per litter are on record, 6 appears to be the usual number. This animal is little studied in the region.

Distribution and Status Small populations of the Grey Wolf remain across Arabia, including a few animals in the Egyptian Sinai (<30), but the Grey Wolf is extinct in the UAE and greatly reduced elsewhere.

Grey Wolves have thick grey or grey-brown fur

Wolves are sociable animals that live in packs

The most substantial numbers remain in Turkey (>5,000), where it still occurs across 75 per cent of its former range, and Iran with >1,000 wolves over about 80 per cent of its former range. Some records indicate that the Grey Wolf once occurred in Egypt and Libya and possibly down the Red Sea Hills to Sudan. Some authorities claim that the Golden Jackal subspecies *C. a. lupaster* is in fact a subspecies of wolf.

Conservation Areas Harrat al-Harrah NR, Hawtat Bani Tamins NR, Raydah Escarpment NR (Saudi Arabia); Jabal Samhan NP (Oman); Judean Desert NR, North Arava NR (Israel); Azraq Oasis WR, Wadi Dana-Finan WR (Jordan); Kavir NP, Golestan NP, Tandourel NP (Iran: occurs in at least 12 other reserves).

Similar Species The Wild Dog (*Lycaon pictus*) once ranged across much of the Sahel, and southern fringes of the Sahara on occasion, from Mauritania to Sudan, ranging north into southern Algeria and possibly other North African countries. The Algerian records were largely associated with the area around the Ahaggar massif. Two wild dogs were shot in 1928 in south-western Algeria. The few surviving populations now occur well to the south, but given their wide-ranging habit occasional sightings are still possible.

SMALL-SPOTTED GENET *Genetta genetta*

Identification This small carnivore has a long, elegant body and tail, with short legs, a relatively long, pointed snout and large, rounded, membranous ears. The tail is thickly haired, with black and white rings along its length. The coat colour is grey to greyish-white, with a liberal scattering of dark to black spots and bars. There are black and white markings on the face.

Size Shoulder height 18 cm.
Weight 1.5–2.6 kg.

Habitat and Behaviour This genet occupies wooded and rocky areas, including those in semi-desert country. Nearly always nocturnal, it lies up in dense cover during the day. A mainly solitary hunter, it takes a wide range of food that includes invertebrates, small mammals (especially rodents), reptiles, amphibians and birds. On occasion it will also eat wild fruits and berries. After a gestation of about 70 days, a litter of 2–5 young each weighing 50–80 g is dropped in a rock crevice or hole, or among dense vegetation. The Small-spotted Genet has not been studied in any detail in the region.

Distribution and Status Some controversy exists as to whether the genet in the region should be called *G. genetta* or *G. felina*. It occurs from northern Western Sahara (Morocco), across the Atlas ranges and coastal plain to north-western Egypt. It also occurs in the Red Sea Hills in Sudan and possibly in the same range in Egypt. It has an extensive sub-Saharan distribution. In the Arabian Peninsula it occurs through the coastal mountains and hills south of Jiddah, Saudi Arabia, extending southwards into the highlands of Yemen. It is also present in the Dhofar of southern Oman. It is almost certain that this population and that in south-west Yemen are linked.

Conservation Areas This species may occur in several conservation areas, but no specific information is available.

Similar Species The Lesser Oriental Civet (*Viverricula indica*) was introduced to the Yemeni island of Socotra, in the Gulf of Aden possibly in the 19th century. It looks similar to a large genet but it is longer in the leg. The Small-spotted Genet does not occur on the island, so there can be no confusion between the two species. The Lesser Oriental Civet's current status on Socotra is unknown, but it was originally brought to the island for the extraction of its anal gland secretion, a base for certain perfumes.

MONGOOSES

LARGE GREY MONGOOSE *Herpestes ichneumon*

Identification As its name implies, this is a large mongoose (total length 1 m and more) with a long, grey-grizzled coat. The tail is prominently black-tipped. The lower parts of the legs are black and short-haired, but over much of the body and tail the hair is long. When the animal is moving, the tail is held curved slightly upwards.

Size Shoulder height 20 cm.
Weight 2.5–4 kg.

Habitat and Behaviour Large Grey Mongooses avoid arid areas and are usually associated with riverine vegetation and other water bodies, although they will wander into adjacent drier habitats when foraging. Rodents make up much of their diet, but they also eat snakes (they will attack and eat even quite large snakes), lizards, amphibians, invertebrates and wild fruits. They deposit their droppings at latrine sites. Commonly seen during the day, they may be solitary or seen in pairs and family parties. In Israel they often live in groups consisting of an adult male, 2–3 females and their young. They frequently stand

on their hind legs to view their surroundings. Normal litter size is 2–4 young, born after a gestation of 75 days. In some parts of this mongoose's range, the pups are born during spring.

Distribution and Status In North Africa this species occurs widely in Morocco, northern Algeria and Tunisia, then there is an apparent range break until north-eastern Libya, across the Mediterranean coast of Egypt including the Nile Delta. It then extends widely down the floodplain of the Nile and beyond, where it has a wide sub-Saharan distribution. In the Middle East it is largely restricted to the Mediterranean coastal plain of Palestine and Israel, Lebanon, northern Iraq and south-east Turkey. It does not occur in Iran.

Conservation Areas Simak Cudi Mountain, Sendinli-Rubaruh (Turkey). Probably occurs in a number through its regional range.

WHITE-TAILED MONGOOSE *Ichneumia albicauda*

Identification This is a large mongoose that stands high on the legs with rump higher than the shoulders and the head carried low. The coarse, shaggy coat is brown-grey to almost black on the legs. The bushy white tail is distinctive and makes up about a third of the animal's total length.

Size Shoulder height 25 cm.
Weight 3.5–5.2 kg.

Habitat and Behaviour These mongooses are found in open woodland, river courses (wadis) that are commonly associated with hill country and coastlines in the vicinity of estuaries. They do not live in true desert areas. In the UAE they are strongly tied to wadis

with permanent water. They eat a wide variety of invertebrates, as well as many small mammals up to the size of hares, and occasionally wild fruits and berries. Within their Arabian range, they enter date plantations and eat fallen fruits. They are nocturnal with some daytime activity, especially during the early morning and late afternoon. Usually solitary animals are encountered. Births apparently peak in the summer months – information is sparse within this region. The usual litter consists of 1–4 cubs.

Distribution and Status The White-tailed Mongoose occurs in and around the Aïr massif, northern Niger and possibly in the Ahaggar region of adjacent southern Algeria. It is found in association with wadis in the Red Sea Hills of south-eastern Egypt and Sudan. It is also present along the coastal plain and wadis cutting into the adjacent mountain ranges of Saudi Arabia, south from Jiddah, Yemen, Oman and the eastern UAE. In Oman it is known from the Batinah coast in the north and an apparently isolated population in Dhofar in the south. It is, however, likely that these populations are linked, because this mongoose utilizes the coastal zone, including intertidal areas, and there are no known barriers to its range.

Conservation Areas None known.

Similar Species Another large mongoose reported to occur in Yemen is the Bushy-tailed Mongoose (*Bdeogale crassicauda*). In 1988 an individual apparently of this species was found near the capital of Yemen, San'a. It appears black at a distance, but at close quarters is more grizzled, although the legs and long-haired, bushy tail are very dark to black. It is known in Africa from only as far north as south-eastern Kenya, favouring open woodland with rocky outcrops.

SMALL INDIAN MONGOOSE *Herpestes auropunctatus*

Identification This, the smallest mongoose in the area, has a limited range. It is identified by its slender build, longish tail and short legs, and fairly coarse, yellowish-grey to olive-grey grizzled coat. The underparts are paler than the upperparts, often pale creamy-buff. The ears are small and are largely covered by hair. There is a rusty-brown ring around the eyes. This is sometimes considered to be a subspecies of the Javan Mongoose (*H. javanicus*).

Size Total length 45–65 cm.
Weight 500–1,000 g.

Habitat and Behaviour In the region the Small Indian Mongoose occupies well-watered and agricultural areas, never venturing into adjacent desert land. It commonly associates with humans, their land and their structures. It is diurnally active and will readily stand on its hind legs to scan an area. It has a varied diet, taking a wide range of invertebrate prey, snakes, lizards and smaller rodents, and will also scavenge at rubbish dumps. The mongoose is generally encouraged around homesteads because it hunts pest rodents and snakes, but

there are numerous reports of it raiding poultry runs and pigeon lofts. Few breeding records are known from the region, but there are indications that births take place in early spring. In Iran these mongooses are said to breed throughout the year, and litters of 2–4 pups are dropped in rock crevices, dense vegetation and even tree holes.

Distribution and Status The species is known from Iraq along the lower Euphrates and Tigris Rivers and the marshes (now very much reduced) at their confluence. It only occurs in southern Iran in the provinces (*ostans*) of Ilam, Bushehr and Seistan. Elsewhere it has a wide distribution eastwards to southern China and South-East Asia.

Conservation Areas None known.

INDIAN GREY MONGOOSE *Herpestes edwardsii*

Identification This species is similar to but larger than the Small Indian Mongoose (*H. auropunctatus*), with a longer, coarser coat. The overall pelage is grizzled grey, with the rusty-brown underfur often showing through on the nape, shoulders, flanks and outer thighs. This colour is also present on the muzzle, cheeks and sides of the neck. The underparts are uniform light orange-brown, the tail tip is off-white to creamy-white and the upper surfaces of the feet are dark reddish-brown.

Size Total length 68–80 cm.
Weight 1–1.5 kg.

Habitat and Behaviour This species favours well-watered, densely vegetated areas such as palm groves and other cultivated land. Like the Small Indian Mongoose, it is commonly found in and around

human settlements. It is a true omnivore that eats anything edible: invertebrates, fish, frogs, reptiles, small rodents and birds; it regularly scavenges and takes some plant food. It can become a pest when individuals take to killing poultry. These mongooses are mainly solitary animals, but pairs and family parties may also be observed, normally during daylight hours. Nothing is known about breeding in the region, but in Pakistan they seem to breed throughout the year.

Distribution and Status Animals in Arabia were probably introduced from Iran, for example on the island of Bahrain, Kuwait and the Saudi Arabian Persian Gulf ports. Individuals are occasionally killed in the UAE, and it is known that these animals are brought in by sailors and labourers. They are not as widespread in Iran as the Small Indian Mongoose. The exact range is not well known, although this species seldom ventures far from the coast. It occurs widely in India and eastwards.

Conservation Areas None known.

MUSTELIDS

LEAST WEASEL *Mustela nivalis*

Identification This tiny aggressive hunter is the smallest carnivore in the region, with a very wide distribution. An elongated, slim, cylindrical body with a short tail (4–6 cm) and short legs distinguishes it from any other species in the region. The coat is short and dense, with upperparts uniform reddish-brown and underparts creamy-white. The upper surfaces of the front feet are white; the hind feet are partially white. In high-altitude areas of Iran, Turkey and possibly

upper reaches of the Moroccan and Algerian Atlas, this weasel moults to white in the winter months. The Egyptian Weasel (*M. subpalmata*) is very similar, but larger.

Size Total length 16–34 cm.
Weight 45–200 g.

Habitat and Behaviour This weasel occupies a wide range of habitats, from coastal to subalpine reaches of high mountain ranges. In many areas it lives around people, buildings and cultivated areas, and it is said to very commonly associate with humans and their dwellings. Weasels are highly agile carnivores that are classic hunters of burrowing rodents, although they do also take other prey including animals larger than themselves. They are active during the day and night. Their home ranges are marked by both males and females; those of the males do not overlap, but a male's range may overlap that of one or more females. This species has not been studied in the region, but elsewhere breeding does not seem to be seasonal. Three to ten pups, with an average of 5 per litter, are dropped after a 35-day gestation. Pups each weigh an average of just 1.5 g at birth.

Distribution and Status This species ocurs across northern Morocco, northern Algeria and Tunisia, as well as north-eastern Libya. It appears to be very rare in Lebanon and may still occur in Israel. It has a wide distribution in Turkey and across northern Iran. In Iran it is also found in the west, where it is closely associated with the Zagros Mountains. The Egyptian Weasel is only known from the lower Nile River and its delta.

Conservation Areas None known.

EUROPEAN POLECAT *Mustela putorius*

Identification Female European Polecats are smaller than the males. Overall coloration is dark brown or black, with pale yellow underfur that clearly shows through. The area between the eye and ear is silvery white, and the lips are white. The domesticated form, known as the Ferret, is very variable in colour. In scientific circles it is believed that the Steppe Polecat (*M. eversmanni*) may also be involved in the Ferret genetic make-up.

Size Total length 50–65 cm.
Weight 400–1,500 g.

Habitat and Behaviour This species has not been studied at all in its very limited North African range. In Europe it most commonly occurs in open forest and meadows. It dens in rock crevices, burrows dug by other species and hollow logs. It is nocturnal and solitary, and hunts small mammals, birds, fish, frogs and invertebrates. Mating takes place in March–June, and after a 42-day gestation a litter of 2–12 pups is dropped, with each pup weighing about 10 g.

Distribution and Status The European Polecat is only known from a limited area of northern Morocco and possibly adjacent areas of Algeria. Tame animals were once kept for rodent control and rabbit hunting, and it is believed that these were the origin of the wild/feral population. The status of the species is unknown in the region.

Conservation Areas None known.

MARBLED POLECAT *Vormela peregusna*

Identification This is a very attractive small carnivore with mottled brown and buffy-yellow upperparts and blackish-brown underparts. A distinctive white band across the face above the eyes contrasts with the largely black to dark brown face. The lip edges are white, as are the upper parts of the rounded ears. The coat has an overall glossy sheen. The fairly short, well-haired tail is usually grizzled whitish with a dark tip. The distinctive coloration and markings are possibly a form of warning to would-be aggressors that if not left alone the animal will spray a noxious-smelling substance from its anal glands. Anal substance spraying is common to many members of the mustelid family.

Size Total length 45–60 cm.
Weight 370–700 g.

Habitat and Behaviour This is a carnivore of open steppe country and low rocky hills. Most of its activity takes place at night, but it is not exclusively nocturnal. It digs its own burrows and will also take over and modify burrows excavated by other species. Although it climbs well it does most of its hunting on the ground. Rodents, birds and reptiles make up the bulk of its prey, but it will also take some invertebrates. Marbled Polecats are solitary foragers with home ranges (about half a square kilometre in Israel and Palestine) that tend to overlap. The mating season in Israel extends from mid-March to early June, with births 8–11 months later, indicating that delayed implantation is taking place. One to eight pups per litter have been recorded.

Distribution and Status This species is absent from Africa and has a very limited distribution in Arabia. It is only known from Israel and Palestine and Lebanon, although there are a few records from Syria and just one from northern Saudi Arabia near the Jordanian border. It occurs in both the European and Asian sectors of Turkey where there is suitable habitat, but it is not clear how widespread it is. In Iran it is widespread in the north from East Azarbaijan in the west to Seistan in the south-east. It occurs in a broad belt beyond the region to eastern China.

Conservation Areas Several in its Iranian range and possibly Israel, otherwise none known.

STRIPED POLECAT *Poecilictis libyca* and *Ictonyx striatus*

Identification Also known as the Saharan Striped Weasel or Libyan Striped Weasel, *Poecilictis libyca* is a distinctive carnivore with conspicuous black and white markings on its face and longish black and white hair on its body (white dominates). Its tail is bushy and mostly white, and the hair on its legs is shorter and black. The similar *Ictonyx striatus* has an extensive sub-Saharan range and may be encountered in the southern range limits of this species in Mauritania and the Red Sea Hills of northern Sudan. The black and white body markings are more clearly defined in *Ictonyx*, and it has a white spot on its forehead. *Poecilictis* usually has a distinctive white band across the face, above the eyes.

Size Total length 34–50 cm.
Weight 200–650 g.

Striped Polecat (Ictonyx striatus)

Habitat and Behaviour The Striped Polecat does not penetrate into the Sahara, but occupies its fringes and adjacent arid zones. Strictly nocturnal, it spends the daylight hours in burrows that it digs itself. It is a poorly known species that eats small rodents, birds, reptiles and a wide range of invertebrates. Like many other mustelids, it can spray a foul-smelling fluid from the anal glands at an attacker. Births are recorded as January–March, with litters of 2–3 young that weigh about 5 g. The gestation period is given in a very wide range, which may indicate that females can employ delayed implantation.

Distribution and Status This species occurs from Senegal and Mauritania in the south-west, across Morocco, Algeria, Tunisia, Libya and Egypt to north-east Sudan. It does not penetrate into the Sahara.

Conservation Areas None known.

STONE MARTEN *Martes foina*

Identification Also known as Beech Martens, these slender and agile carnivores, with their longish bushy tails (about 25 cm) and prominent ears, should not be mistaken for any other species, except perhaps for the Pine Marten (*M. martes*) where ranges overlap. Their coat is soft and uniformly light grey-brown, with a strong brown tint on the rump that is much denser and has a distinct gloss in winter. The summer coat is more drab. There is an irregular off-white to white throat patch.

Size Total length 85–110 cm.
Weight 1–2 kg.

Habitat and Behaviour This marten occupies a wide range of habitats, from woodland to open rocky country, as well as agricultural areas. Like other mustelids it eats a variety of food, including animal and plant foods. In the region, Stone Martens are known to den up in rock crevices, boulder clusters, burrows excavated by other species, hollow trees and even buildings. Most activity takes place at night, but they will move at dusk and dawn in undisturbed areas. Male home ranges are larger than those of females. In parts of the range mating takes place in midsummer, but fertilized eggs are not implanted at that time and births are delayed until the following spring. Litters of 3–4 young are the norm.

Distribution and Status The Stone Marten does not occur in Africa, and in Arabia it is limited to Israel and Palestine, Lebanon and parts of Syria and Iraq. Populations occur through Turkey and Iran. It has been fairly heavily hunted for its pelt, resulting in local extinctions and population reductions.

Conservation Areas None known.

PINE MARTEN *Martes martes*

Identification The Pine Marten is similar to the Stone Marten, but has more luxuriant winter fur and a light yellow throat patch. The two species can also be separated based on habitat choice.

Size Total length 60–95 cm.
Weight 800–1,800 g.

Habitat and Behaviour This is a species of both coniferous and deciduous forests, well adapted to an arboreal way of life. Each adult has several nest sites, usually in tree holes, scattered through fairly large home ranges. Male home ranges are larger and may overlap those of one or more females. Pine Martens are nocturnal and range considerable distances each night when hunting. They take a wide variety of both animal foods, up to the size of squirrels, and plant foods. Mating takes place in midsummer, but the young are not born until the following spring because of delayed implantation. Normal litter size is 3–4 young, but as many as 8 have been recorded.

Distribution and Status Because of its luxuriant winter coat, this species has been heavily hunted. In the region it is only known from the Caspian forests associated with the Alborz Mountains, where it occurs in very low numbers. It is present in Turkey, but its status there is unknown.

Conservation Areas None known.

HONEY BADGER *Mellivora capensis*

Identification One of two badger species in the region, the Honey Badger or Ratel is easy to identify by its size and distinctive coloration. It is a thick-set carnivore that is stocky with a short, bushy tail and very small ears. Its upperparts are silvery-grey (in some animals the mantle is almost white). Its underparts, legs and upper foot surfaces are black, and its hair often has a sheen. The upper and lower body coloration is separated by a white to off-white line that runs from the top of the head and on to the tail. The tail is often held erect while walking. The paws are large and the front feet have powerful long claws.

Size Shoulder height 30 cm.
Weight 8–14 kg.

Habitat and Behaviour The Honey Badger occurs in a great variety of habitats, although it is absent from the heart of true desert country. It takes a very wide range of food, including invertebrates, rodents, reptiles (including large snakes), birds, carrion and wild fruits. Much of its prey is obtained by digging. Its common name is derived from its tendency to break into both wild and domesticated bee hives to eat the honey and larvae. It is this habit that sometimes brings it into conflict with humans. Honey Badgers also scavenge around rubbish dumps on settlement outskirts. Most hunting takes place on the ground, but they also climb well to raid birds' nests. If left alone they pose no threat, but if disturbed or angered they are tough and aggressive.

Honey Badgers are usually seen singly or in pairs, or when young accompany a female. Most activity takes place at night, but where they are not disturbed they will also forage during the day, especially during the mornings and late afternoons. They will dig their own burrows, but also take over burrows dug by other species, hidden up among rocks or in dense vegetation. Little is known about their

breeding habits in the region. Elsewhere they tend to be non-seasonal, with a female producing 1–4 cubs (usually 1–2), of which often only one is successfully raised. The gestation period is about 180 days. In Turkmenistan, bordering north-eastern Iran, young are known to be born in April–May.

Distribution and Status The Honey Badger is not common anywhere in the region, occurring on the fringes of the Sahara in the south from the Atlantic to the Red Sea, then in the west across Mauritania into Morocco as far as the Atlas ranges and to west-central Algeria. It is present in south-east Egypt and southwards through much of Africa. It has a wide but sparse distribution in Arabia, across northern, south-west and eastern Iran. It has not been recorded in Turkey.

Conservation Areas None known.

EURASIAN BADGER *Meles meles*

Identification This species is superficially similar to the Honey Badger in build, with greyish grizzled upperparts and very dark to black underparts and legs. The black and white facial stripes are distinctive, as is the short, light-coloured tail. Animals tend to be heavier in autumn than they are in spring.

Size Shoulder height 65–80 cm.
Weight 10–34 kg.

Habitat and Behaviour The Eurasian Badger shows a preference for forest and densely vegetated terrain, and is commonly found in association with cultivated areas. It is omnivorous, taking a very wide

range of animal and plant foods. It digs for much of its food with the long, powerful claws on its front feet. Badgers excavate large and often complex burrow systems (setts), frequently among tree roots and often on sloping ground. Unlike the Honey Badger, the Eurasian Badger is social, with family groups, called clans, sharing a sett. It is strictly nocturnal, but will lie outside den entrances during the day. The sleeping sections of the sett are lined with dry plant material. During severe weather, such as snowstorms, Eurasian Badgers may remain in their dens for weeks and even months. It is not known to what extent this applies in the region. Births of 2–6 cubs usually take place in February–March across much of this badger's range. Young badgers each weigh just 70 g at birth.

Distribution and Status This species is absent from North Africa and most of Arabia, but known to occur in Israel and Lebanon, with a few records from Iraq and Syria. It is relatively widely distributed in Turkey and occurs in the forests of northern Iran, as well as the south-west, including suitable areas in the Zagros ranges. Elsewhere it lives from Western Europe to China and Siberia.

Conservation Areas None known.

EURASIAN OTTER *Lutra lutra*

Identification This is the only otter likely to be encountered in the region. The overall body coloration is dark brown, looking almost black when wet, and the lips, cheeks, throat and upper chest are white to fawnish-white. The feet are webbed and each toe has a distinct claw. The muscular tail is thick at the base and tapers towards the tip; it is longer than the head and body length (50–82 cm).

Size Shoulder height 30 cm.
Weight 6–10 kg.

Habitat and Behaviour Eurasian Otters occupy a wide range of inland waters, such as rivers, lakes, dams and marshes, as well as estuaries and sheltered areas of coastal waters. On occasion they temporarily wander away from water. Although they are mainly fish eaters, they also hunt frogs, crabs, insects, birds and small mammals. Adult males patrol, mark and defend a territory within which one or more females and their dependent young are resident. They communicate with whistling calls. Otters construct burrows, called holts, at the edges of water bodies; the entrances may be above or below the surface level. Otters also shelter among rocks and in dense vegetation. Usually 2–3 cubs weighing about 130 g each are dropped after a 59–63 day gestation. In Israel and Palestine otters are noted to breed at any time of year; in Armenia births have been recorded in May.

Distribution and Status This species is heavily hunted for its pelt over much of its range, and many populations have been greatly reduced. Pollution and modification of water systems are also responsible for its decline. In North Africa these otters are strongly tied to rivers in and to the north of the Atlas ranges in Morocco, Algeria and Tunisia. They are present in Israel and Palestine, Lebanon, Jordan, Iraq, and possibly Syria, Turkey and Iran. In Iran they are known from the Caspian region and other areas, but absent along the south coast and the central desert region.

Conservation Areas En Nimfit NR (Israel). It is probably found in other areas, but information about these is not available.

INDIAN SMOOTH-COATED OTTER *Lutrogale perspicillata*

Identification This species is very similar to the Eurasian Otter (*Lutra lutra*), except that its tail is strongly dorso-ventrally flattened.

Size Shoulder height 30 cm.
Weight 7–11 kg.

Habitat and Behaviour The Indian Smooth-coated Otter favours flatland waters and commonly enters the sea. It is also said to hunt on land, sometimes quite far from water.

Distribution and Status In the region the Indian Smooth-coated Otter is only known from the lower Tigris River and especially the marshes near its mouth. The Iraqi marshes have been largely destroyed, so if this species survives it will probably only do so in the marsh area that spills into Khuzestan in south-west Iran. This is an apparently isolated population – the species is otherwise known only from eastern Pakistan to South-East Asia.

BEARS

BROWN BEAR *Ursus arctos*

Identification This is the largest surviving carnivore in the region. It is powerfully built with a very short tail, and has a large head with a longish muzzle and prominent rounded ears. The legs are fairly long and heavy, with large paws armed with long claws. The coat is fairly long, shaggy and coarse, with its colour ranging from pale sandy-yellow or grey to darker brown.

Size Shoulder height 90–120 cm.
Weight 120–250 kg.

Habitat and Behaviour The Brown Bear probably once lived in a wide range of habitats in the region, but is now restricted to mountain country (although at least in Iran it will move onto the plains at times). Brown Bears are true omnivores that take a wide range of animal and plant foods. They will kill larger mammals, including domestic livestock up to the size of a horse, and can become a nuisance. Brown Bears hibernate in the mountain ranges where snow falls in winter, such as the Zagros and Alborz in Iran. Apart from digging dens they will also lie up in caves, crevices and dense vegetation. In the region they are mainly solitary except when a female is accompanied by young. Mating takes place in May–July over much of the bear's range, but implantation is usually delayed until October or November. Cubs are born during January–March and usually number 2 (with a range of 1–4), each weighing more than 300 g.

Distribution and Status The Brown Bear has disappeared from many areas across its regional range. Populations used to occur in Palestine, Lebanon, Jordan and Syria, but the species is now extinct in these countries. Small numbers may survive in the Kurdish region of Iraq. Scattered populations persist across Iran, particularly in the Alborz and Zagros mountain ranges. The species is similarly scattered across Turkey in mountainous and forested areas. In both these countries the numbers of bears have been greatly reduced. Brown Bears were said to have survived in the Atlas ranges of North Africa until about the middle of the 19th century. Little is known of this extinct population – it is possible that animals noted in historical times were introduced, although they are known there from the recent fossil record.

Conservation Areas Kiamaki WR, Arasbaran PA, Lisar PA, Parvor PA, Khosh-Yeilagh WR, Galestan NP, Sarani PA, Tandoureh NP, Oshtrankuh PA, Dasht-e Arjan PA (Iran).

ASIATIC BLACK BEAR *Ursus thibetanus*

Identification Smaller than the Brown Bear, as its name implies the Asiatic Black Bear has a glossy black coat, a brown muzzle and a white or yellowish, V-shaped patch on the chest.

Size Shoulder height 70–80 cm.
Weight 65–150 kg.

Habitat and Behaviour The Asiatic Black Bear favours dense woodland and thicket in hill and mountain country. Very little is known about it in the region, but it is said to be nocturnal and to readily climb trees to reach the variety of fruits that make up a large part of its diet. It also takes animal foods.

Distribution and Status The Asiatic Black Bear is restricted to far south-eastern Iran within the province of Baluchestan (eastern Makran Range), where it is considered seriously endangered.

Conservation Areas Possibly in the Gându (Bâhukalât) PA (Iran).

INTRODUCED CARNIVORES

RACOON *Procyon lotor*

Identification The Racoon is larger than a domestic cat, with quite long, greyish fur, and a bushy tail that is about a third of the animal's total length and is ringed with black bands. The face is sharply pointed, with distinctive black and white markings, and prominent erect ears.

Size Shoulder height 22–30 cm.
Weight 2–12 kg.

Distribution and Status The Racoon is native to North and Central America and was introduced to Europe to be farmed for its fur. Animals escaped and populations have established themselves. Racoons from fur farms in the former USSR (Russia) became established in the Caucasus and gradually spread southwards. They were discovered in the Iranian province of Gilan, on the south-west of the Caspian Sea, in 2002, and may have arrived in the area at least ten years earlier. An animal of forest and woodland, the Racoon can be expected to spread widely in northern and western Iran.

RACOON DOG *Nyctereutes procyonoides*

Identification The Racoon Dog sports a long coat, especially in winter; this ranges in colour from grey and yellowish to reddish. The legs, feet and chest are dark. The tail is only about a third of the animal's total length and is covered with thick hair. The species looks especially dumpy in autumn and winter, when it carries fat reserves. Its legs are short, and it has a distinctive black facial mask, small, rounded ears and a pointed muzzle.

Size Shoulder height 20 cm.
Weight 4–10 kg.

Distribution and Status This wild dog from the Far East was introduced to fur farms in Europe, from which animals escaped. It is now well established and its range is spreading. It has not yet been recorded in the region, but it is spreading down the western shore of

Racoon

Racoon dog

the Caspian Sea – following the same route as the Racoon (*Procyon lotor*) – and can be expected to reach Iran and possibly Turkey. On the western shore of the Black Sea it has already become established in Bulgaria, and it is likely to reach European Turkey in the near future.

PRIMATES

Our closest relatives, primates have large, complex brains, the ability to use their hands for locomotion and feeding, and in most cases highly social behaviour.

BARBARY MACAQUE *Macaca sylvanus*

Identification This is the only primate in Africa north of the Sahara. It is relatively large (males are larger than females) and no tail is visible. The coat is thick and yellowish-grey to grey-brown. The face is naked and dark flesh-coloured.

Size Shoulder height 40–50 cm.
Weight 6–13 kg.

Habitat and Behaviour The Barbary Macaque is mainly confined to rugged, isolated, scrub-covered mountain slopes, gorges and montane forests dominated by cedar and oak trees. Barbary Macaques eat a wide variety of plants. Cedars are of particular importance in the harsh winter months. Troops may include several adult males and 7–40 individuals. Home range sizes for each troop vary from as little as 25 ha to as much as 1,200 ha. A single young weighing 400 g is born in February–June after a gestation of 210 days.

Distribution and Status Previously more widely spread in North Africa, this macaque is now found only in isolated locations in northern Morocco and Algeria. As much as 75 per cent of the former population probably persists in the Middle Atlas range in Morocco. Smaller populations are located in the High Atlas and Rif (Morocco), and Petite Kabylie and Grande Kabylie (Algeria). It is likely that less than 20,000 individuals survive.

Conservation Areas Talassemtane R, Toubkal NP (Morocco); Djurdjura NP, Taza NP, Babor R, Chrea NP, Gouraya NP (Algeria).

HAMADRYAS BABOON *Papio hamadryas*

Male Hamadryas Baboon (right) is larger than female

Identification The Hamadryas Baboon is also called the Sacred Baboon because of its religious significance to the ancient Egyptians. Males can be as much as twice as large as females. Adult males have well-developed, silvery-grey capes extending down the back and on to the shoulders and chest. The hair on the sides of the head is also long. Females are browner and lack the cape. The facial and buttock skin is reddish-pink. Hamadryas Baboons have long, dog-like snouts.

Size Shoulder height 40–60 cm.
Weight 10–20 kg.

Habitat and Behaviour These baboons occupy areas of arid and rocky hill country, but rarely at great altitude. Seeds, roots and bulbs make up a large part of their diet. In the vicinity of towns and cities in their Arabian range they regularly scavenge at rubbish dumps. Hamadryas Baboons live in a hierarchial group system (harem), with a single male and 1–3 females, and dependent young. Several groups may join together in bands of up to 60 individuals. They circulate within a home range, sleeping on steep cliffs or rocky outcrops, always within reach of drinking water. Females probably give birth at any time of the year to a single young after a 175-day gestation.

Distribution and Status This baboon is restricted to the Horn of Africa, eastern Ethiopia, Eritrea, the Red Sea Hills of Sudan and the mountain ranges and hills of western Saudi Arabia and Yemen. In Saudi Arabia it occurs from about the vicinity of Medina south-westwards to Jiddah and then throughout into Yemen, where it is known as far eastwards as Wadi Idim in the Hadramaut.

Conservation Areas Raydah Escarpment NR (Saudi Arabia).

Similar Species Other primates that may be encountered in the northern Sahel belt on the fringes of the Sahara include the Savannah or Olive Baboon (*Papio cynocephalus*) – shown on page 93 – and the Patas Monkey (*Erythrocebus patas*). Isolated Olive Baboon populations apparently occur in the Aïr-Ténéré massif in northern Niger to the south of the Algerian border and the Tibesti massif on the border between Chad and Libya, as well as in the Ennedi ranges in north-eastern Chad close to the Sudanese border. Patas Monkeys are large, principally terrestrial primates that favour open savannahs and rocky areas, in which (for monkeys) they occupy very large home ranges. An apparently isolated population may have been recorded in north-western Niger.

Olive Baboons occur in montane areas edging the Sahara

AARDVARK (Tubulidentata)

This is an unusual mammal that is the only member of its order. It is a highly specialized ant and termite hunter.

Aardvarks are very distinctive in appearance

AARDVARK *Orycteropus afer*

Identification The Aardvark's large size, long, pig-like snout, long, tubular ears, kangaroo-like tail, stout legs and hunched back make it unmistakable.

Size Shoulder height 60 cm.
Weight 40–70 kg.

Habitat and Behaviour Aardvarks show a strong preference for grass and scrub plains, but can be expected wherever their principal prey – ants and termites – is abundant. The powerful front claws of Aardvarks are used for tearing into ant and termite colonies, and the ants are then extracted with the sticky tongue. Aardvarks are mainly nocturnal and solitary, spending the day in deep and often extensive burrow systems. There may be several burrows within a home range. Occupied burrows usually have numerous small flies in their entrances. A single 2-kg young is born after a 210-day gestation.

Distribution and Status The Aardvark occurs along the Red Sea coast into Sudan, and in the Sahel belt to southern Mauritania. It can apparently be found as far north as the Aïr massif in Niger, close to the Algerian border. Early Egyptian paintings of this species indicate that it might once have occurred as far north as the Mediterranean.

Conservation Areas None known.

HEDGEHOGS AND SHREWS (Eulipotyphla)

The Eulipotyphla order includes the hedgehogs and shrews. All have high metabolic rates and eat voraciously in order to maintain themselves.

HEDGEHOGS

Six species of hedgehog occur across the region. All have a covering of short, hard spines on their upperparts and when threatened roll themselves into a defensive ball to protect their underparts and head. The spines are alternately banded brown-black and white. The tail is very short and often not visible. Hedgehogs have a total length of 15–32 cm and a weight of 250–1,600 g.

Hedgehog species differ mainly in coloration, habitat choice and distribution. All are nocturnal and usually solitary, and in some parts of their range they hibernate. They eat a wide range of animal foods and fruits and fungi. Two to ten young, each weighing 8–18 g, are born after a 35–48-day gestation.

Ethiopian Hedgehog (*Paraechinus aethiopicus*) Sometimes called the Desert Hedgehog, the muzzle of this species is dark brown, contrasting with the rest of its face, which is white; it has fairly long ears. This species occurs across North Africa and most of Arabia, but is absent from Turkey. It has not been reliably reported from Iran, but is present on several Persian Gulf islands.

White-bellied Hedgehog (*Atelerix albiventris*) The underparts and face of this hedgehog are very pale to white. It is a species of the African tropics, but occurs across the Sahel and southern Sahara fringe in northern Niger.

Algerian Hedgehog (*Erinaceus algirus*) This species is very similar to the White-bellied Hedgehog, but occurs north of the Sahara from Morocco and into Libya.

Long-eared Hedgehog (*Hemiechinus auritus*) This is a small species with long ears and white to fawnish face and underparts. It is found on the coastal plain of eastern Libya and Egypt, and also occurs in Israel and Palestine, Lebanon, Jordan, Syria, Iraq, the east coast of Saudi Arabia, Bahrain and possibly the western UAE.

Brandt's Hedgehog (*Paraechinus hypomelas*) This large-eared, dark brown to black hedgehog is the darkest hedgehog in the region. It is found mainly in the mountains and hills of south-west Saudi Arabia, Yemen, northern Oman and the eastern UAE, as well as in central Iran eastwards to Baluchestan.

East European Hedgehog (*Erinaceus concolor*) The overall colour of this hedgehog is pale to darkish dull brown, and it has short ears. It favours agricultural and wooded areas in Turkey, north Iraq and north-west Iran, and is also found in Israel and Palestine, Lebanon, Jordan and Syria.

Ethiopian Hedgehog

White-bellied Hedgehog

Algerian Hedgehog

Long-eared Hedgehog

Brandt's Hedgehog

East European Hedgehog

SHREWS

Shrews are small, short-legged, mouse-like mammals, with long, wedge-shaped snouts, very small eyes and short ears. All eat insects and other invertebrates (and occasionally small vertebrates). They may be active day and night. At times they reach quite high densities, but they are mainly solitary creatures.

For their size, shrews are very aggressive and at least some species vigorously defend territories. They occupy many different habitats, but most species are associated with damp and moist locations. In desert areas they may occupy oases, wadi bottoms and mountainous terrain with pools. Breeding of these interesting insectivores has been little studied in the region, but many are probably not seasonal breeders, especially at lower altitudes. Gestation periods vary between 17 and 28 days; litters range from 2 to 10 pups, which are sheltered in a nest of dry plant material.

How many species of shrew occur across the region is a topic of debate, but conservatively there are at least 22. The majority belong to the genus *Crocidura* (white-toothed shrews) with 17 species, *Sorex* (long-tailed shrews) with 2 species and *Neomys* (water shrews) with 1 species. Although difficult to identify to species level by layperson and scientist alike, all shrews are easily recognizable as shrews.

Savi's Pygmy Shrew (*Suncus etruscus*) This is one of the two smallest mammals in the world, weighing just 2.5 g with a length of about 7 cm. It has a wide distribution in the region.

House Shrew (*Suncus murinus*) This is one of the largest shrews in the region, with a length of up to 24 cm and a weight of up to 100 g in some parts of its range (in the region about 30 g). It was probably introduced accidentally from southern Asia and is now commonly associated with coastal towns in Egypt, around the Arabian Peninsula and Iran. It freely enters houses, as suggested by its common name.

A white-toothed shrew (Crocidura sp.)

SENGIS/ELEPHANT SHREWS (Macroscelidea)

Of the 17 species of sengi, only one occurs in the region. In recent years the name sengi has been favoured because although elephant shrew is descriptive of the animals' long, trunk-like snouts, they are in fact related to neither elephants nor shrews.

NORTH AFRICAN SENGI *Elephantulus rozeti*

Identification This is the only sengi in the region, with a range from Morocco (including north Western Sahara), across northern Algeria and Tunisia, to western Libya. If seen up close it should not be mistaken for any other species within its range. It has an elongated, cylindrical, trunk-like snout that is constantly twitching, large, membranous ears and prominent eyes. The hind legs and feet are considerably longer than the front ones, and the tail is about half of the total length. The upperparts are buffy in colour and the underparts are white to off-white. The coat is very soft.

Size Total length 19–23 cm.
Weight 40–50 g.

Habitat and Behaviour This species is found in rocky areas, where it shelters in crevices, and it apparently uses burrows dug by other species on gravel plains. It is nocturnal and crepuscular, but frequently sun basks near its shelter. Sengis feed mainly on insects and other invertebrates. For their size they can move at great speed, giving the impression of a 'low-flying puff-ball'.

Distribution and Status The North African Sengi occupies a mixture of habitats from sea level to the Atlas plateaux. It is common in many locations.

PIKAS, RABBITS AND HARES (Lagomorpha)

The order Lagomorpha includes pikas, rabbits and hares, which are characterized by their highly developed hind limbs and long ears (except pikas). Both of these characteristics are particularly noticeable in the hares.

RUFESCENT PIKA *Ochotona rufescens*

Identification There is only one species of pika in the region. This has long, dense, soft fur that is greyish-brown and usually has a reddish tinge. There is no visible tail, short, rounded ears and short legs. The feet are well furred underneath.

Size Total length 18 cm.
Weight 180 g.

Habitat and Behaviour Pikas live mostly in rocky areas, but in parts of their range Rufescent Pikas live on open plains. In the former case they live in crevices, and otherwise they dig burrows. They are mainly diurnal, and even in winter when snow falls they do not hibernate. They feed on a wide range of plants and are well known for the 'haystacks' they construct near their shelters. These piles of plants are harvested to ensure that they have enough food to get them through the coldest months. In some areas pikas occur at high densities (up to 70 per ha in northern Pakistan). There are two litters each year, in spring and summer, each with 4–10 pups.

Distribution and Status The Rufescent Pika occurs in the region only in Iran, where it has a wide distribution that includes the Alborz and Zagros ranges.

The North American Ochotona princeps – a similar pika species

EUROPEAN RABBIT *Oryctolagus cuniculus*

Identification The European Rabbit is the ancestor of domestic rabbits and has the same form, with longish ears (shorter than those of hares). The fur is soft and a grizzled grey-brown, the nape is reddish-buff and the short tail is white below and brown-black above. The underparts are buffy-white and the undersides of the feet are haired.

Size Total length 39–50 cm.
Weight 1.3–2.2 kg.

Habitat and Behaviour European Rabbits show a preference for sandy soils in rolling to hilly terrain, often with patches of good plant cover. They dig complex burrow systems or warrens, and live in colonies. They are mainly nocturnal, but not infrequently seen during the day as they often bask in the sun. A female may have several litters in a season, each with 5–6 (up to 9) pups.

Distribution and Status This species occurs on the Mediterranean coastal plain and adjacent hills from Morocco to Tunisia. Some authorities believe that it occurs here naturally, but others think it may have been introduced from Spain.

HARES *Lepus* spp.

Three hare species are recognized across the region, although they involve taxonomic controversies. All are similar in appearance, with very long ears, short (7–14 cm) black and white tail, very long back legs and soft, dense fur. The overall coloration is variable, but the upperparts are grizzled grey-brown and in desert areas often very pale; the underparts are at least partly white, and a reddish-buffy nape patch is often present. Distribution is often the best identification criterion for species recognition. These species have a total body length of 45–70 cm and weigh 1.2–4.5 kg.

European Hare

Cape Hare

All species of hare are mainly grass eaters, but they also consume herbaceous plants, especially in the most arid areas. They are mainly night active and solitary, but at favoured or productive feeding sites several may be seen together. Unlike rabbits, hares do not burrow but lie up in bush or grass cover, the weight of their body creating a 'form'. In the region all three species may breed throughout the year, but in some areas breeding may be seasonal and females may have multiple litters of 1–4 pups each.

Cape Hare (*L. capensis*) In desert areas during the day, the Cape Hare may dig a shallow burrow no deeper than the length of its body. Cape Hares tend to run into the open when disturbed; African Savannah Hares make straight for cover. This species occurs right across North Africa, the Arabian Peninsula and all of the Middle East except Turkey. It favours open habitats often with minimal cover and sparse vegetation. It tends to thrive in areas of overgrazing such as much of the Sahel.

African Savannah Hare (*L. victoriae*) This hare overlaps with the Cape Hare in the Sahel and on the Atlantic coast to as far north as Western Sahara (Morocco). It may occur as far north as western Algeria, where an apparently isolated population has been located. It favours more scrub-covered and hilly country than the Cape Hare, but the two species may be seen close to each other. Some authorities believe this species to be the same as the Scrub Hare (*L. saxatilis*).

European Hare (*L. europaeus*) This species is sometimes lumped with the Cape Hare. The European species tends to be darker overall. In the region it appears to be restricted to Turkey and north-western Iran. It requires open country with thickets or other cover, and frequents agricultural land.

RODENTS (Rodentia)

This very large and diverse group of mammals is well represented in the region. Rodents are the only mammals with a large, prominent pair of chisel-like incisor teeth at the front of the upper and lower jaw.

ASIAN PORCUPINE *Hystrix indica*
NORTH AFRICAN PORCUPINE *Hystrix cristata*

Identification These two porcupines are very similar and cannot be mistaken for any other species. The upper body is covered with long, black-and-white banded quills and spines, and a crest of very long, coarse hair extending from the top of the head to the shoulders. The quills, spines and hair are all raised when the animal is alarmed. The rest of body is covered in black hair, with some animals having a pale to white crescent on the throat. The short tail (14 cm) has a cluster of hollow 'rattle quills'. The Asian Porcupine is usually smaller in the Arabian Peninsula than it is elsewhere, weighing 9–10 kg, although larger individuals are known.

Size Shoulder height 25 cm.
Weight Asian Porcupine 10–24 kg; North African Porcupine 8–22 kg.

Habitat and Behaviour Porcupines occupy a wide variety of habitats, but are absent from true desert. In many areas they show a strong preference for rocky country. They are nocturnal, and spend the day in burrows or caves, or among rocks. Although they are mainly solitary foragers, pairs and family groups are common and several animals may share burrows and home ranges. They feed on a wide variety of plants, digging for bulbs, roots and corms, and eat tree bark. Litters consist of 1–2 well-developed young.

Distribution and Status The North African Porcupine occurs across the Sahel and northwards through western Mauritania, Morocco (including Western Sahara), northern Algeria and the Libyan coastal plain. It used to occur in Egypt, but is believed to be extinct there. The Asian Porcupine lives around the edges of the Arabian Peninsula, but

Although porcupines are solitary foragers, they are social

Quills and spines are formidable defence weapons

is absent from northern Oman and the UAE, as well as the Persian Gulf coast. It occurs in Israel and Palestine, Jordan, Syria, eastern Turkey and Iraq, and is widespread in Iran, but absent from deserts and high mountains. It is heavily hunted throughout the region as an agricultural pest and for its meat.

GUNDIS (Ctenodactylidae)

Three species of gundi occur in the region and all are rather like guinea pigs in appearance. They have short but long-haired tails, soft and sleek coats, small, rounded ears and fairly large eyes. The overall coloration is similar in all species, with a range of browns and buffy-yellows. Gundis have a length of 17–29 cm and weigh 160–290 g.

Gundis occupy desert and generally dry habitats, where they live in rock crevices from sea level to 2,400 m (7,874 ft). They forage for a variety of plant foods during the day, and also frequently bask in the sun. They live in small colonies that may number 3–11 individuals. One to three young are born fully haired, with eyes open, and are soon able to move around. Litters appear to be dropped in January–June, but this varies from area to area.

The Long-haired Gundi (*Massoutiera mzabi*) occupies the Saharan massifs, including Ahaggar, Tassili n'Ajjar, Tibesti and possibly Aïr. Gundis *Ctenodactylus gundi* and *C. vali* are found in Morocco through to north-western Libya. There are two more species that occur on the fringes of the region: Speke's Pectinator (*Pectinator spekei*) in eastern Ethiopia, Djibouti and Somalia, and the Senegal Gundi (*Felovia vae*) in a small area of adjoining Mauritania, Senegal and Mali.

The gundi Ctenodactylus gundi

GROUND SQUIRRELS

Fulvous Suslik (*Spermophilus fulvus*) This species is found only in northern Iran. It has a total length of 28 cm, of which the short tail is just 4 cm in length. The overall coat colour is pale grey. The animal occupies flat areas with fine silt or clay soils in which it excavates burrows. Little is known of this species.

Long-clawed Ground Squirrel (*Spermophilopsis leptodactylus*) With a total length of 34 cm, this species is quite stocky and has a short, well-haired tail that is partly black below and white above. The coat is sandy to greyish-yellow in colour, and the winter coat is particularly silky. This species lives in the sandy desert country in the Sarakhs area of north-eastern Iran in small family groups in burrows. It is diurnal and eats a wide variety of plants and some insects. Three to six pups are dropped in April–May.

Barbary Ground Squirrel (*Atlantoxerus getulus*)

With a total length of 34–45 cm and a weight of 325 g, this squirrel has a longish coat with a white stripe along each side and sometimes a white stripe down the centre of the back. The underparts are white and the tail is quite bushy. It occupies the Atlas ranges in Morocco and extends marginally into Algeria, from sea level to 4,000 m (13,123 ft). Normally inhabiting rocky areas, it digs burrows, lives in colonies and is diurnal. It eats mostly tree fruits and seeds.

Western Ground Squirrel (*Xerus erythropus*) This species occurs from East Africa to the Atlantic (in Mauritania), in a belt that incorporates

much of the Sahel. An apparently isolated relict population occurs in Morocco inland from the town of Agadir. It is larger than the Barbary, has a fairly long, bushy tail, fairly short, coarse coat with white lateral stripes and very short ears. The underparts are white. It digs burrows, lives in colonies and is diurnal.

PERSIAN SQUIRREL *Sciurus anomalus*

Identification The upperparts of this tree squirrel are usually dark in colour (in some areas they are lighter), and the front part of the body and tail may be reddish (also the underparts). Some populations are grey above, yellow below and have a reddish tail. The tail is about a third of the total length.

Size Total length 35 cm.

Habitat and Behaviour This is a diurnal rodent that feeds on the seeds of cedar, pines and oaks, and also tree buds and leaves. The seeds are stored in tree hollows for the winter.

Distribution and Status The Persian Squirrel lives in Israel, Lebanon, Syria, eastern Iraq, eastern Turkey and where oak forest survives in the Zagros mountains of Iran. It is considered to be rare and endangered in many areas.

Conservation Areas Mt Hermon NR (Israel) and Dibbin Forest NP (Jordan).

Similar Species In Baluchestan, far south-eastern Iran, the Palm Squirrel (*Funambulus pennantii*) lives in date plantations and wooded areas. It has distinctive dark and white stripes on its back.

JERBOAS (Dipodidae)

Ten species of jerboa of the genera *Dipus*, *Jaculus* and *Allactaga* occur in the region, across North Africa, the Arabian Peninsula, and northwards into eastern Turkey, and they are widespread in Iran. They vary in detail, such as the number of functional toes on the back feet

Lesser Jerboa (Jaculus jaculus)

and the length of the ears, and in size, but they are all recognizable as jerboas. Their hind legs and feet are exceptionally long, and they move with a rapid hopping gait. The tail is very long, with a distinctive tuft of longish black and white hair at the tip, and this is used for balance when hopping. The ears range from moderately long to long. The fur is soft and fawn to yellowish to reddish above, and the underparts are normally white. Total length is 25–36 cm; weight is 55–140 g.

Jerboas are most commonly associated with arid areas, including sandy desert, but a few species – such as the Euphrates Jerboa (*Allactaga euphratica*) – occupy areas with higher rainfall. Their activity is restricted to night-time. They dig their own burrow and several may live in close proximity. During the hot summer months burrow entrances are closed during the day. The bulk of jerboas' food is made up of seeds and succulent plants. Some species breed throughout the year, while others are more seasonal. There may be 2–9 pups in a litter (this varies between species).

GERBILS *Gerbillus* spp.

The large group composing the *Gerbillus* genus consists of mostly small mice with long tails that usually have a tuft of longer hair at the tip. Many species are adapted to living in desert or semi-desert areas, and one or more species occur in many locations in the region. All are similar in overall appearance and differ largely in the detail. At least 30 species occur across North Africa and the Middle East: many are widespread, and a few are very localized. Total length is 12–30 cm; weight is 10–63 g.

Gerbils have very soft coats that range in colour from grey to red-brown and underparts that are pale to white. The back feet are quite long, but never as long as those of the jerboas. Gerbils move on all four feet, the soles of which are hairy. All species are mostly nocturnal and dig their own burrows. Most live in loose colonies; others are solitary. Gerbils eat a variety of plant foods, including seeds, roots and grasses, and at times insects (especially termites) make up an important part of their

Cheesman's Gerbil (Gerbillus cheesmani)

diet. Breeding in most species is poorly known, but the majority seem to breed at any time of the year. Gestation is 20–22 days, litter size is 1–8 (usually 4–5) and most young weigh little more than 2 g at birth.

A few species are widespread, including the Pygmy Gerbil (*Gerbillus henleyi*), which ranges from Algeria to Egypt and into the Arabian Peninsula. The Baluchistan Gerbil (*G. nanus*) occupies the area from Algeria eastwards across much of Iran. Cheesman's Gerbil (*G. cheesmani*) ranges across the Arabian Peninsula and southern Iran.

FAT-TAILED GERBIL *Pachyuromys duprasi*

Identification This is a very distinctive species with a short, thickened and club-like tail (4–6 cm). It has very soft, fine, yellowish-grey to buffy-brown fur and white underparts. There is a white spot behind each ear.

Size Total length 16 cm.

Habitat and Behaviour The Fat-tailed Gerbil lives on gravel plains (*hamada*) with patches of low vegetation. Its main food is insects, but it probably includes some plant foods in its diet. It is thought to breed throughout the year. Litters comprise 3–5 pups.

Distribution and Status This species is distributed across the northern Sahara from Morocco to Egypt.

JIRDS *Meriones* spp.

At least 11 species of jird occur across the region; some are very widespread while others are localized. Most species are larger than the biggest gerbils. Their tails are well haired, with longer hair towards the tip forming either a tuft or a crest. The coat colour varies from yellowish-buff and fawn to various shades of brown. The underparts are always paler, and in a number of species they are white. Jirds are more rat-like than gerbils, and have narrow ears. Their tails are about half of their total length, which is 19–38 cm. They weigh 30–100 g.

Jirds occupy a wide variety of habitats that include sand and gravel desert, grassland and cultivated areas. All dig their own burrows, which can be very long and complex; within these there are lined nest chambers as well as food-storage areas. Some jirds live in colonies of varying density, while others have limited social contact. Some species are seasonal breeders; others may have young at any time of the year. Gestation varies between 20 and 30 days, litters number 1–9, and a female may have several litters in a season. Two of the most widespread species across North Africa and the Middle East are the Libyan Jird (*Meriones libycus*) and Sundevall's Jird (*M. crassus*).

FAT SAND RAT *Psammomys obesus*

Identification This rather stout species has a fully haired, thick tail with a distinct black tuft of long hair towards the tip. The overall colour varies from reddish-brown to sandy-buff, with paler underparts.

Size Total length 24–33 cm.

Habitat and Behaviour This species' habitat choice seems to be limited by the availability of the succulent plants that provide much of its food in desert areas, whether these be sandy or rocky. It digs complicated burrow systems and lives in colonies. In areas where it occurs at high densities it is easy to observe because it is active during the day. Fat Sand Rats probably breed throughout the year, and litter size is usually 2–5 pups.

Distribution and Status The species occurs widely across much of the Sahara and its fringes from Algeria to Sudan and north-western Arabia, and there are scattered populations on the outer fringes of Saudi Arabia.

Sundevall's Jird (Meriones crassus)

Fat Sand Rat

OTHER MICE AND RATS

BROWN OR NORWAY RAT *Rattus norvegicus*
BLACK OR HOUSE RAT *Rattus rattus*

Identification These are large rats, the more heavily built Brown Rat especially, with long tails. The Brown Rat's tail is shorter than the Black Rat's, and it has proportionally shorter ears. The coat colour in both species is extremely variable.

Size Total length Brown Rat 40 cm; Black Rat 37 cm.
Weight Brown Rat 300 g; Black Rat 150 g.

The Brown Rat (shown here) is larger than the Black Rat

Black Rat

Habitat and Behaviour Brown and Black Rats thrive in towns, villages, farmsteads and on agricultural land, and reach their highest densities around and near coastlines. In some towns in Arabia, they show little fear and forage freely among rubbish and in gardens. Both species appear to be dependent on access to drinking water, which limits their spread into waterless areas. The Black Rat is more widespread in isolated settlements. Rats are omnivores that take a wide range of food. They can breed throughout the year, with each female having several litters and 5–10 pups per litter.

Distribution and Status Both these species are exotic to the region, but now have a very wide distribution and are commensal with man.

Similar Species The Short-tailed Bandicoot Rat (*Nesokia indica*) could be mistaken for these two rats, but its tail is shorter than its head and body, its ears are much smaller, and its overall body hair tends to be longer and variable in colour. Its head appears shorter and blunter. It rarely enters buildings and is associated with rivers, irrigation canals and other permanent waters. Bandicoot Rats dig extensive burrow systems and feed on a wide range of plant foods. They can cause considerable damage to agricultural crops, hence their alternative name of Pest Rat. They occur from the Lower Nile and delta, eastwards to Israel, Iraq, Syria and wherever there is suitable habitat in Iran.

SPINY MICE *Acomys* spp.

Three species of spiny mouse occur in the region, although some authorities believe that many more species should be accepted. They are all easily recognized because of their typical mouse-like appearance and backs covered with sharp, inflexible spines. Total length is 11–30 cm; weight is 11–90 g.

Spiny mice live in cracks and crevices in rocks, and eat seeds and other plant material, as well as invertebrates including snails. Accumulated food debris is a sure sign that spiny mice are in residence. In some places they have become commensal with man. They probably breed throughout the year, and have litter sizes of 1–5 young that are often born with their eyes open and are weaned after 2 weeks.

The Golden Spiny Mouse (*Acomys russatus*) occurs from eastern Egypt to Israel and Palestine, with scattered populations in the Arabian Peninsula. One other species has a limited range in south-central Turkey. The one most likely to be encountered is the Egyptian Spiny Mouse (*A. cahirinus*), which ranges across much of North Africa (though probably not in the Sahara proper), and widely in Arabia and southern Iran.

Egyptian Spiny Mouse (Acomys cahirinus)

BARBARY STRIPED GRASS MOUSE *Lemniscomys barbarus*

Identification This and other members of the *Lemniscomys* genus are sometimes called zebra mice. Their upperparts are brown to dark brown, and have numerous pale stripes that extend from the shoulders to the rump on either side of a dark stripe running down the centre of the back. They should not be mistaken for any other species in the region.

Size Total length 19–24 cm.

Habitat and Behaviour This species constructs small, round nests of grass and leaves near the ground, with well-defined paths radiating from them. A range of seeds, green plant material and sometimes insects is eaten. Breeding seems to be tied to seasonal rains, and litters usually have 4–5 pups, each weighing about 3 g.

Distribution and Status This mouse occurs in the dry savannah and plains that circle the Sahara, from Morocco to Tunisia and the Sahel to the south.

NILE OR KUSU RAT *Arvicanthis niloticus*

Identification This is a stocky rat with a tail that is shorter than the head and body length, smallish rounded ears and short hairs almost covering the tail skin. The ears are small and rounded, and the hair on the back is reddish-brown. The coat is rough and varies in colour from buffy-grey with a brownish tinge to a richer brown. The underparts are paler and usually greyish-white, with the upper foot surfaces reddish-brown, like the snout.

Size Total length 25–30 cm.

Habitat and Behaviour The Nile Rat occupies a range of habitats, including dry rocky hillsides with vegetation tangles and grass. It digs long burrows. Observations in Dhofar show that it also lives among boulders. It is both day and night active, and in some areas is easy to observe. Nile Rats are mixed plant feeders that take a range of seeds and succulents.

Distribution and Status This species is quite widespread in eastern Egypt and the vicinity of the Nile and its delta. In Arabia it was until recently only known from south-western Yemen, but it also occurs in the Dhofar of southern Oman. It is colonial and may reach high densities, which can be in pest proportions when populations increase.

Barbary Striped Grass Mouse

Nile or Kusu Rat

BATS (Chiroptera)

Bats are the only mammals capable of powered flight. They are divided into two major groups: the fruit-eating bats (Megachiroptera) and the insect-eating bats (Microchiroptera). The fruit-eating bats are generally much larger in size than those in the insect-eating group. Only two species of fruit-eating bat occur in the region, but there are at least 58 species (21 genera) of insect-eating bat. Obviously all of them cannot be covered here, so this section includes just a few of the most common species you may encounter in the area.

EGYPTIAN FRUIT BAT *Rousettus aegyptiaca*

Identification This large bat has a pointed, dog-like head, large eyes and a very short tail. The fur of the upperparts is dark brown and the underparts are grey-brown, sometimes with a brighter brown tinge.

Size Wingspan 60 cm.

Habitat and Behaviour The Egyptian Fruit Bat is one of the very few fruit bats that can navigate by echolocation. It spends the daylight hours roosting in caves, in colonies that can number from just a few hundred to several thousand. The species feeds on a wide range of ripe wild and cultivated fruits, including dates and figs. It also seeks out nectar, and in this way serves as a pollinator for a number of tree species. For the first six weeks of its life the single young is carried by the mother when she goes off to forage. After this, when it is too heavy, it is left at the roost until it can forage on its own.

Distribution and Status This bat occurs along the Gulf of Oman and Persian Gulf coastal plain of Iran, around the coast of the Arabian Peninsula from the UAE, Oman, Yemen and Saudi Arabia. It is also found in parts of Turkey, the eastern Mediterranean coastline and adjacent interior, and westwards into Egypt, especially along the Nile.

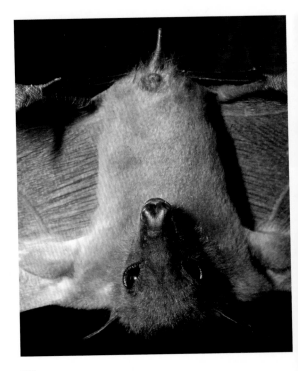

STRAW-COLOURED FRUIT BAT *Eidolon helvum*

Identification Its large size, dog-like face and yellowish straw-coloured fur make this species unmistakable. An orange neck collar is often present in males.

Size Wingspan 75 cm.

Habitat and Behaviour Although this is mainly a species of the African tropics, it is recorded fairly regularly in western Yemen and adjacent south-western Saudi Arabia. It is a highly migratory species, and it is not clear whether this population is resident or whether it moves into the area as part of the dispersal that takes place from the Congo Basin and its fringes. In Africa it forms colonies numbering hundreds of thousands, and even millions. Records indicate that the largest number counted in a Yemeni roost was 20.

Distribution and Status This species is found in western Yemen and south-western Saudi Arabia, and has been recorded along the Nile River.

Straw-coloured Fruit Bats at roost

Straw-coloured Fruit Bats

MUSCAT MOUSE-TAILED BAT *Rhinopoma muscatellum*

Identification This is one of three very similar bats in the region: two have a wide distribution, but this species is known only from Iran, Oman and the UAE. It is easy to identify by its long, mouse-like tail that projects well beyond the membrane. A membrane connects the ears across the forehead.

Size Wingspan 30 cm.

Habitat and Behaviour These bats are mainly cave dwellers, but also make use of old, dark and abandoned buildings. They form small colonies, but individuals roost in a well-spaced manner, clinging to rock walls. The main birth season appears to be July–August, but this is not known for certain.

Distribution and Status This species is found in Iran, Oman and the UAE.

EGYPTIAN SLIT-FACED BAT *Nycteris thebaica*

Identification Although this is not the only long-eared (3.4 cm) bat in the region, this species has a shallow-forked tail-tip and a long-lobed slit down the centre of the face that is unique to this group of bats. The fur of this species is very variable: it is usually buffy-brown, but in desert areas it may be greyish-white. The underparts are always paler.

Size Wingspan 24 cm.

Habitat and Behaviour This species has a very wide habitat tolerance and is not fussy about roosting sites. Roosts may contain anything from a few individuals to several hundred. Unlike many bats that feed in flight, this species has fixed feeding sites to which it returns with its insect prey, usually moths, beetles and grasshoppers, some of which is snatched off the ground. A litter of inedible fragments accumulates below the perch.

Distribution and Status The species occurs across North Africa and mainly in the western sector of the Arabian Peninsula, but also in Riyadh in the east.

HORSESHOE BATS *Rhinolophus* spp.

Seven species of horseshoe bat are known in the region, and all are similar in overall appearance and require detailed examination to bring them to species level. They all have a distinctive horseshoe-shaped main nose-leaf that is situated above the upper lip. Two skin outgrowths (the sella and lancet) are also located on the face. The outer edge of the tail membrane is more or less squared-off and not pointed towards the tip.

One or more species of horseshoe bat occurs in virtually all parts of the region. One of the most widespread is Cretzschmar's, Bokhara or Geoffroy's Horseshoe Bat (*Rhinolophus clivosus*), which roosts in caves, mine shafts and old buildings.

Horseshoe bats enfold their bodies in their wings at roost

EGYPTIAN FREE-TAILED BAT *Tadarida aegyptiaca*

Identification This is one of five species of free-tailed bat that occur in the region, all of which are very similar. As their name implies, they are characterized by a tail that is enclosed for only a third to half its length in the tail membrane, with the remainder extending beyond the edge of the membrane. The tail is always shorter than the tails of the mouse-tailed bats. Also known as bulldog or mastiff bats, free-tailed bats have short heads with heavy wrinkling on the upper lip. Their coat colour varies from greyish-brown to dark brown on the back and slightly paler underneath. The hair is short and very soft.

Size Wingspan 30 cm.

Habitat and Behaviour Free-tailed bats occupy a great variety of habitats, and frequently roost in the roofs of buildings. The bats may roost in small numbers, but sometimes do so in the hundreds tightly packed together. The colonies give off a rubbery smell, especially on hot and wet days. Free-tailed bats are one of the few types of bat that scuttle around on the ground when moving within the roost and occasionally when hunting. The authors have seen them snatching up termites while poised at the edges of these insects' emergence burrows. Otherwise, they are high and fast fliers taking insects on the wing.

Distribution and Status Free-tailed bats are known from Algeria to Egypt, widely southwards, patchily around the edges of the Arabian Peninsula and across southern Iran.

GLOSSARY

Blaze Usually a light-coloured (sometimes used for dark) mark down the front of the face.

Browser Feeding mainly on woody or herbaceous plants.

Canine A single tooth situated immediately behind the incisors in each jaw; usually tall and pointed, used as part of the killing apparatus in many carnivores.

Carnivore An animal that preys on other animals for its food (predator); a member of the order Carnivora.

Commensal Animals such as rats and mice living in close association with humans, often in their structures.

Crepuscular Active during the twilight hours of dusk and dawn.

Diurnal Active during the daylight hours.

Echolocation While they are in flight, insectivorous bats emit high-frequency sound waves that help them to avoid collision with obstacles and to locate prey.

Endemic Found only in a particular region, and nowhere else in the world.

Exotic Not native to a country or region, but introduced from other countries or areas. Also 'alien'.

Feral Having reverted to a wild state.

Foraging Searching for or seeking out food.

Gestation The period between conception and birth in which offspring are carried in the uterus.

Grazer Any organism eating mostly grass.

Herbivore An animal that feeds mainly on plants.

Hibernate To pass part of or the whole winter in a dormant state with the metabolism decreased.

Home range The area covered by an animal in the course of its day-to-day activities.

Incisors Sharp-edged front teeth, usually in both the upper and lower jaws.

Insectivore A mammal that eats mostly insects.

Introduced species Animal or plant brought by humans from areas where it occurs naturally to areas where it has not previously occurred. Some introductions are accidental, others deliberate.

Nocturnal Active at night.

Omnivore An organism that feeds on both plant and animal foods.

Reintroduced animal Where an animal has become locally or regionally extinct, and individuals of the same species are returned from other populations.

Rut Period of sexual excitement (in male animals) associated with the mating season.

Scavenger An animal that feeds on dead or decaying organic matter.

Species A group of interbreeding individuals of common ancestry, reproductively isolated from all other groups.

Territory A restricted area inhabited by an animal, often for breeding purposes, and actively defended against other individuals of the same species.

Wadi Term in Arabic-speaking countries for a valley, ravine or channel that is dry except in the rainy season.

PICTURE CREDITS

Principle photography by Chris and Tilde Stuart.

Additional photography:
Koen de Smet 19 (Slender-horned Gazelle).
James Dolan 27 (Barbary Red Deer).
I.R. Iran DoE-CACP/WCS/UNDP (through Luke Hunter) 51 (Iranian Cheetah).
Marijcke Jongbloed 48, 49 (Striped Hyaena).
André Mader 76 (Indian Grey Mongoose); 111 (Fat Sand Rat).
John Newby, Sahara Conservation Fund 65 (Pale Fox).
Walter Poduschka 97, centre (Algerian Hedgehog); 97, bottom, right (East European Hedgehog).
Galen Rathbun 99 (North African Sengi).
Klaus Rudloff 55, 56 (Jungle Cat); 58 (Eurasian Lynx); 68 (Golden Jackal); 77 (Least Weasel); 79 (Marbled Polecat); 85 (Eurasian Badger); 89 (Black Bear); 104 (Asian Porcupine); 105 (Gundi); 106 (Barbary Ground Squirrel and Western Ground Squirrel).
J. Usher-Smith (through Christian Gross) 37 (Arabian Tahr).
Roland Wirth 20 (Cuvier's Gazelle).
Iain Green/NHPA 75; Manfred Danegger/NHPA 82; John Shaw/NHPA 100; James Warwick/NHPA 87.

ACKNOWLEDGEMENTS

It is not possible to acknowledge everyone who has assisted us over the years, but a few stand out. Christian Gross was extremely helpful during our earlier Arabian involvement. The founder of the Arabian Leopard Trust, Marijcke Jongbloed, was our first 'pathway' into the Arabian Peninsula. Moaz Sawaf served well as field assistant and translater. Koen de Smet (ANB, Belgium), Klaus Rudloff (Berlin Zoo, East), Galen Rathbun, Marijcke Jongbloed, Andre Mader, John Newby, Bill Houston, Luke Hunter (Wildlife Conservation Society, NY) and Farid Belbachir (Universite de Bejaia, Algeria) are thanked for helping to fill our photographic gaps. At New Holland, James Parry is thanked for embracing the idea, which was seen through its various stages by Krystyna Mayer and Simon Papps. The authorities and rulers in a number of countries allowed us to work over the years in the regions under their control. Many people offered help, advice and hospitality over the years that we have been active in research and survey work in the area covered by this guidebook.

FURTHER READING

Firouz, E., 2005, *The Complete Fauna of Iran*, I.B. Tauris, London.
Haltenorth, T. and H. Diller, 1984, *A Field Guide to the Mammals of Africa including Madagascar*, Collins, London.
Harrison, D.L. and P.J.J. Bates, 1991, *The Mammals of Arabia*, Harrison Zoological Museum, Sevenoaks, England.
Nowak, R.M. (ed.), 1999, *Walker's Mammals of the World*, 2 volumes, John Hopkins University Press, Baltimore.
Stuart, C. and T. Stuart, 2006, *Field Guide to the Larger Mammals of Africa*, Struik, Cape Town.

INDEX